THE BERRY BOOK

Also by Robert Hendrickson:

THE GRAND EMPORIUMS
THE GREAT AMERICAN CHEWING GUM BOOK
THE GREAT AMERICAN TOMATO BOOK
RIPOFFS
LEWD FOOD
HUMAN WORDS

THE BERRY BOOK

The Illustrated Home Gardener's Guide to Growing and Using Over 50 Kinds and 500 Varieties of Berries

by Robert Hendrickson

Doubleday & Company, Inc.,
Garden City, New York
1981

To my mother-in-law, Florence, and in memory of my father-in-law, Salvatore Maggio.

Library of Congress Cataloging in Publication Data

Hendrickson, Robert, 1933–
The berry book.

Includes index.
1. Berries. 2. Cookery (Berries) I. Title.
SB381.H53 634.7
ISBN 0-385-13589-0
Library of Congress Catalog Card Number 78–22326

Interior designed by Virginia M. Soulé

Printed in the United States of America
First Edition

Contents

Introduction

No other country has a tradition of berry growing and gathering equal to ours in America, where we have led the world in developing plump, juicy berries and produced berry-pie makers par excellence. Unfortunately, this tradition has been waning in recent years with the paving of so much of our country and the freezing of so much of our food. It is my intention in this book to help revive a nationwide interest in berry growing, which is surely one of the most satisfying and rewarding forms of home gardening.

Home-grown berries can be raised by anyone with a small patch of ground in the backyard (or by city gardeners on a terrace)—they are *not* specialty items for expert gardeners alone, as so many people believe. Additionally, berries can be planted in space-saving ways in the back, front, and on the sides of the house, and they cost practically nothing to care for. These delicious fruits often aren't obtainable anywhere unless they are home-grown, or if available, can cost as much as four dollars a pint to buy. And if you have a large enough space, you can earn money raising fresh berries for the local market, as in the case of precious, perishable raspberries, which bring anywhere from one to four dollars a pint, depending on the area and season. A mere quarter-acre of raspberries, for instance, should yield at least 500 quarts, grossing a bare minimum of one thousand dollars.

In these pages you'll find hundreds of berries, ranging from those used to make bathtub gin to those eaten as breath-sweeteners. If I had stuck to the strict botanical definition of what a berry is, many of these berries would have been omitted. Technically, a berry is a fleshy fruit that doesn't usually split open, has few or many seeds (but no "stone" like a cherry), and always develops from one enlarged ovary. This definition immediately eliminates "berries" I have covered like the strawberry, blackberry, and raspberry, and includes unlikely true berries such as the eggplant, tomato, and grape. What I have obviously done here has been to treat only small fruits that are commonly considered berries, no matter what the botanists say. In the case of the tomato, I have included a few odd species that taste more like berries than tomatoes, but I entirely omitted the grape, believing that vineculture cannot really be done justice in any space briefer than an entire book.

Despite these restrictions, *The Berry Book* covers over fifty species of berries that can be grown in the home garden and gathered in the wild, not to mention some five hundred varieties developed from these species. Rather than impose any doctrinaire theory on the reader, I have tried to cover all methods of growing berries—allowing gardeners to choose what works best for them—and I have included all aspects of growing from purchasing and setting out plants to collecting berries in the wild and propagating them at no cost. Close to one hundred berry recipes have also been provided—recipes for soap, candles, and love potions among them. So, with many thanks to all the authorities, witches, and warlocks who helped me and without any further ado, on to the cultivation and lore of red strawberries and *white* strawberries, of raspberries and blueberries and blackberries . . . of candleberries and soapberries . . . of candy berries and laxative berries . . . and even of those berries that give us bathtub gin . . .

—Far Rockaway, New York
October 6, 1980

THE BERRY BOOK

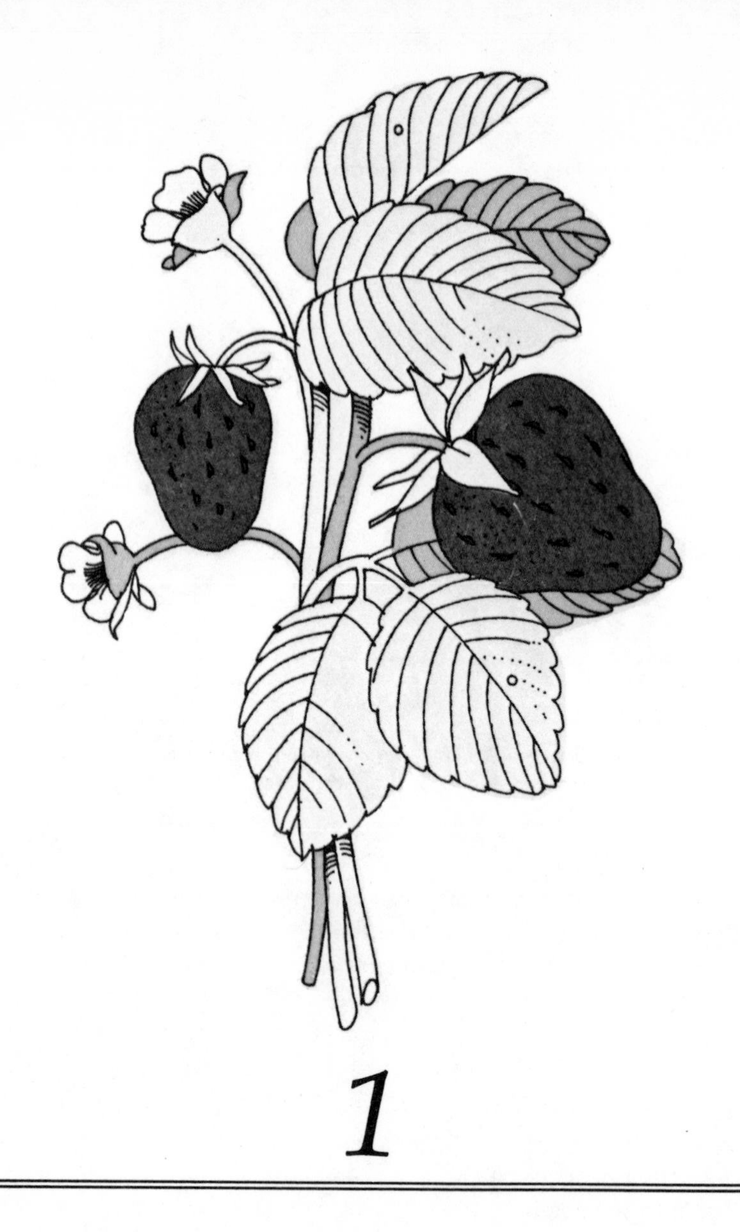

1

Strawberry Fields Forever

"Doubtless God could have made a better berry, but doubtless God never did," William Butler remarked about the strawberry and Izaak Walton promptly quoted the sentiment in *The Compleat Angler,* immortalizing it for posterity. Butler was given to excesses—he started the old fishwives' tale that it's dangerous to eat oysters during months without an "R" in them—yet concerning the strawberry the old gastronome hardly exaggerated. Not only is this member of the rose family "the very primrose of fruits, announcing the season has opened again," as another gardener put it, but it is more widely esteemed for its flavor than any other fruit, and it is the most commonly grown fruit in the home garden. Strawberries will grow almost everywhere throughout the world, need little room, multiply themselves each year by producing "daughter plants," have no thorns as so many small fruits do, need no trellises and supports, and can produce good crops without spraying because they bear their fruit before pests get a good start in the garden. About their only defect, in the words of Dr. George M. Darrow, former chief horticulturist of the USDA, is that "the modern American back does not seem to have adapted itself to strawberry picking."

The romantic history of the strawberry goes back to the Greeks and Romans. The early Greeks had a taboo against eating any red foods, including wild strawberries, and this added mystery to the fruit, leading many to believe it possessed great powers. Though mentioned in the writings of Virgil, Ovid, and Pliny, the plant began to be cultivated only in the Middle Ages, so far as is known. Some wild beliefs were associated with it. Pregnant women, for example, avoided strawberries because they believed their children would be born with strawberry marks (small, slightly raised birthmarks resembling strawberries) if they partook of the fruit. On the other hand, strawberries were considered to be a medicinal cure for almost everything; it was even thought that a lotion made of their roots could fasten loose teeth by strengthening the gums.

The strawberry of the Middle Ages, often portrayed in Gothic art, was the little wild, or woods, strawberry, the *fraise des bois* (*Fragaria vesca*) of the French, that sends out no runners and is much esteemed for its taste. The gardeners of Henry VIII collected roots of this type from the woods to plant in the royal gardens, and it is said that Cardinal Wolsey introduced the English to wild strawberries and cream. Strawberry leaves symbolize the rank of a duke, which indicates how highly the berries were regarded. Several theories have been proposed about the origin of the name *strawberry* itself, but none is convincing. Some say the straw mulch often used in its cultivation inspired the name, some hold that the achenes with which the berry's surface is dotted resemble straw, others claim that the dried berries were once strung on strands of straw for decorations, while still others say that the long, strawlike runners of the mother plant (strawlike when dry) gave the fruit its name. The mysterious word was used as early as A.D. 1000 in England and doesn't derive from any other language. It may be that the *straw* in *strawberry* is a corruption of the word *strew*. Certainly the mother plant strews or scatters new plants all over a patch when it propagates itself by sending out runners, and the fruit is strewn among the leaves on the plant itself.

In America, Indians collected and even cultivated the North American wild strawberry, *Fragaria virginiana,* which is much bigger than its wild European counterpart. Roger Williams, who founded Providence, Rhode Island, in 1636, wrote that the strawberry was "the wonder of all the Fruits growing naturally in these parts. The Indians bruise them in a Morter, and mixe them with meale and make strawberry bread." A New Amsterdam settler wrote that "the flat land near the river is covered with strawberries, which grow so plentifully in the fields that one can lie down and eat them." For their part, the Indians liked the berries simply sweetened "with the dew of milkweed." But the colonists, to their everlasting credit, invented that old American favorite strawberry shortcake, the largest of which is baked today at the annual Lebanon, Oregon, strawberry festival, where a huge cake towering over twelve feet in the air is cut with a two-man saw every year.

A real-life espionage thriller marked the beginning of the strawberry as we know it today. As early as 1624 the larger wild American strawberry had been sent to France for cross-breeding with the European wild strawberry. But real progress in producing the big modern varieties began when Captain Amédé Frézier, a French explorer-spy, was observing Spanish fortifications on the west coast of South America and noticed the giant-fruited but tough Chilean wild beach strawberry, *Fragaria chiloensis,* which native Indians had been cultivating for centuries. On completing his mission, Frézier dug up five plants and brought them back to France. He claimed that they bore fruits "as large as walnuts," but no one could tell at first, for although he was an amateur botanist who should have known better, the captain had selected only female plants with no males to pollinate them. This was remedied by planting the Chilean species next to American types, and eventually the big pineapple strawberry (*Fragaria ananassa*) was crossbred from the two.

All strawberries came to be named Fraisiers in French in honor of Frézier, but another Frenchman, botanist Antoine Nicholas Duchesne, had more to do with the fruit's development. Duchesne, in 1764, at the age of seventeen, presented Louis XV with a potted plant laden with ripe strawberries, winning favor with the king, and his book *L'Histoire des Fraisiers* later became a classic. Among Europeans only the eccentric English horticulturist Thomas Andrew Knight surpassed him in strawberry expertise. Knight was the first grower to practice large-scale, systematic strawberry breeding, and his work with the pineapple strawberry marked the beginnings of the big, sweet strawberries grown today.

Many Americans had a hand in developing the modern strawberry. Our earliest breeder of note was bearded Charles Mason Hovey, a Cambridge, Massachusetts, nurseryman and publisher of a horticultural magazine who in 1834 originated the first fruit variety cross of any kind made by an American. Another pioneer was New Yorker James Wilson, whose 1851 Wilson strawberry variety enabled commercial acreage to jump from a few thousand to over 100,000 acres within thirty years. Others who contributed much to native strawberry development included growers as varied as a drygoods merchant, a Louisiana railroad station agent, a housewife, and a mountain homesteader. Henry A. Wallace, America's thirty-third Vice-President, was a respected experimental grower who developed a great many delicious varieties.

It is a tribute to American ingenuity that of the 130,000 acres devoted to commercial strawberries here, a mere 8,000 acres in California planted with the Shasta and Lassen varieties account for a quarter of the world's crop. But mass-produced strawberries, though they are being bred more and more for flavor, are usually varieties that are grown because they are "firm" (read "tough") and ship well. They are also picked before they ripen, and while strawberries do gain color off the plant, they cannot develop any more flavor after being picked and invariably lose in vitamin content after two days. Many growers, too, aren't above putting a few choice berries atop a box and filling the rest with small or rotten berries. This practice, incidentally, dates back at least to 1624 when Francis Bacon wrote of "the strawberry wives, that laid two or three great strawberries at the mouth of the pot, and all the rest were little ones." Almost four centuries have passed and the strawberry buyer still must beware!

The best answer is still to grow strawberries at home, as Americans have been doing since the 1700s. Types like Florida 90 (the berry usually found in stores and restaurants during the winter) are good enough when you can't grow your own, but why compromise with quality when it isn't necessary? A Suwannee or a Fairfax, or almost any variety left to ripen in the sun, is so much juicier and sweeter than a commercial berry (even those bought at roadside stands) that it tastes like another fruit entirely, and as a bonus has stored up much more vitamin C because it was stimulated by sunlight longer. Grow strawberries in patches, in borders, or on the front lawn, grow them high up on a terrace in a strawberry barrel or in a strawberry pyramid, try them in the greenhouse, in a common planter, or in an old-fashioned strawberry jar, even experiment with so-called "climbing types." (There are descriptions of all these techniques further on.) But plant some strawberries this spring. Twenty-five plants—which can produce from 1 to 5 quarts of berries apiece, depending on how well they are grown—should be plenty for a family of four and could last a lifetime if you keep developing their offspring. On the other hand, some growers like to plant a 25-foot row of strawberry plants for each member of the family, so that there's enough for preserves and freezing. If you want to make some pin money growing the berries, figure it this way. Commercial growers in California, using all the latest methods, harvest 50,000 quarts of berries from an acre. Do the same in a 33 × 66-foot home strawberry patch, which is exactly one-twentieth of an acre, and you can harvest 2,500 quarts. Sell these at an average seventy cents a pint (home berries should be worth even more) and you've made $3,500. Not bad for a little plot of land and no more than a few days' work, even if you cut these figures in half.

The Best Strawberry Varieties To Grow

Some thirty-five species of wild strawberry grow around the world (see Wild Strawberries, in Index), but all cultivated strawberry varieties originated from crosses between the three wild species previously mentioned and a second wild European species, the Hautbois strawberry (*Fragaria moschata*), a close relative of *Fragaria vesca* or the woods strawberry. Literally thousands of varieties have been developed from these four species. Strawberries don't grow in the sea any more than red herrings grow in the wood, but they do grow almost everywhere else. There are strawberry varieties available for practically every area in the United States, you can choose early and late types, and there are novelties galore—including "everbearers," "climbers," and even a *white* strawberry.

Following are several lists to help you select the strawberry best suited for your needs, including a main list of about 100 early, midsummer, late, and everbearing varieties. Be sure to purchase virus-free (VF) plants when buying any of these—almost all nurseries offer them—and save yourself lots of trouble. All these varieties are "perfect flowered" and do not require another variety to be planted nearby for pollination purposes. You can tell perfect-flowered varieties when they bloom by the center of the flower, which will consist of a rounded mass of light green petals surrounded by large yellow anthers, full of pollen, at the ends of the stamens; "imperfect" or pistillate varieties will have only pistils, no anthers or pollen.

*Strawberry Superlatives**

BEST FLAVOR— Suwannee, Fairfax, Fortune, Honeylump, Fletcher, Midland

LARGEST STRAWBERRIES— Robinson, Jerseybelle, Midland, Catskill, Titan

EARLIEST BEARING— Earlidawn

BEST LATE BEARING— Redstar, Vesper, Marlate, Late Giant, Harvest Stretcher

†*"WILDEST" FLAVOR*— Dunlap, Ogallala, Late Giant

MOST UNUSUAL— Snow King (White), Black Prince (Blackish)‡

BEST "EVERBEARING"— Ozark Beauty (heaviest bearing), Ogallala (best-tasting)

MOST DISEASE-RESISTANT— Guardian, Surecrop, Quinault, Delight, Redchief, Howard, Sunrise, Apollo

HARDIEST— Chief Bemidje, Dunlap, Trumpeter, Ogallala, Fort Laramie

LONGEST SEASON— New Empire

* See main variety list for more details about each type.

† Flavor closest to that of wild strawberries.

‡ An old English variety that isn't available from any nurseries in the U.S., but which might be tracked down by writing to nurserymen in England.

Everbearers

Despite the fact that nursery catalogs rave about them, "everbearing" strawberries are still a subject of controversy among the experts. Everbearing varieties fruit more or less continuously throughout the summer and fall until frost, some varieties fruiting on their runners and others only fruiting on the mother plants. In the Southeast all types grown might be considered everbearers, since they produce several crops of fruit; on the other hand, where the growing season is very short, "everbearers" will only bear once. Professor John P. Tomkins of Cornell University claims that "most growers would be better off to have a good June-bearing variety only, produce their fruit over a two-week period and store it in the freezer for the remainder of the growing season." Some gardeners instead pinch all early blossoms off everbearers and grow them only for their fall crop. Still others complain that everbearers bear too sparsely, the crop produced over so long a period that only small pickings can be made in a single day. But most pomologists seem to favor everbearers if the right variety is used. Professor George Darrow, author of *The Strawberry,* a fascinating technical work with lots of information for the general reader, favors the everbearer, but qualifies his enthusiasm. "In general," he writes, "everbearers are high-flavored and plant hardy—hardier than non-everbearing varieties. Because strawberries are shallow-rooted and subject to drought and because seed do not set in very hot periods in summer [seed are necessary for berry development], everbearing strawberries do best in deep rich soil in cool climates. In the southern states periods of hot weather stop fruit setting and development for much of the summer so that everbearers are not generally successful there." He recommends everbearers only for southern New Jersey west to northern Missouri and northward, but many gardeners claim to have success with them in the South using varieties like Ozark Beauty.

Following is a list of the best everbearing varieties, according to the USDA and other experts. See the main variety list for more details.

1. Ozark Beauty
2. Geneva*
3. Ogallala
4. Red Rich
5. Streamliner
6. Quinault
7. Mastodon
8. Fort Laramie
9. Chief Bemidje
10. Arapahoe
11. Honeylump
12. Sugarball
13. Superfection
14. Progresser
15. Rockhill
16. Gem

* Both Geneva and Red Rich have Fairfax, which many regard as America's best-tasting strawberry, as one parent.

"Climbing" Strawberries

"'Climbing strawberries,' are a fallacy," says Professor George L. Slate of the New York State Experiment Station. "Strawberry plants do not climb." Professor Slate goes on to explain that all strawberries produce runners, that nurseries selling "climbing strawberries" simply sell vigorous everbearing strawberry varieties that send out a lot of runners and instruct the home gardener to tie the runners to a trellis. This has many disadvantages. "If the runner plants are attached to the trellis," Dr. Slate says, "they do not produce roots as do the plants that are prostrate on the ground. The whole trellised system of plants is forced to absorb nutrients and water supply through the root system of the one mother plant, a most inefficient arrangement . . . Also, runner plants on a trellis will die during the winter as it is not feasible to protect them. If they are removed from the trellis and laid on the ground, they wither and die or are so severely winter-injured, they are worthless. Strawberry plants must have a well-developed root system in the soil to live through the winter."

Nevertheless, despite such expert testimony, home gardeners continue to buy novelty "climbing" strawberries—and some have even reported success with them, this perhaps an example of mind over matter. If you decide to try them, certainly don't pay the dollar or so a plant many nurseries are asking. Just look for any everbearer that produces fruit on its runners in the main variety list following and order these, tying the runners to a trellis as they grow. I can't guarantee a "five-foot wall of berries" as one nursery does, but at least your disappointment will be less expensive.

*Old-fashioned Strawberries**

Dunlap (1890)	Brooklyn Scarlet
Fairfax (1923)	Hovey
Suwannee	Wilson (1840)
Harvard 17 (1909)	British Queen
Midland (1929)	Loxford Hall Seedling
Lady of the Lake	President
Forest Rose	Sir Joseph Paxton
Dr. Hogg	

* These old-fashioned varieties were described by gardening writers as outstanding in flavor as long as a century ago. As far as I have been able to determine, only the first five (see main variety list) are offered by either American or English nurseries, but the adventurous might want to hunt for them by writing nurserymen in the U.S. and abroad trying to track them down. The first eleven are American varieties, the rest British. All of the undated berries above were being cultivated before 1890.

Varieties for Home Freezing

Sparkle*	Pocahontas
Fletcher*	Ozark Beauty
Frontenac*	Earlidawn
Midland	Apollo
Redglow*	Northwest
Surecrop*	Catskill
Tennessee Beauty	Marshall
Badgerglo	Cyclone
Midway	Hood

* Indicates varieties tested and recommended by USDA; see main variety list for more details. Twenty-four quarts of strawberries (one crate) will yield 38 pints of frozen berries.

*Foreign Favorites**

Hampshire Maid (English, everbearer)
Royal Sovereign (English, early)
Cambridge Aristocrat (English, midseason)
Cambridge Favourite (English, early)
Hummi Gento (everbearer)
Ostara (everbearer)
May-Queen (English, early)
La France (early)
General Chanzy (early)
Profusion (English, early)
Belrubi (midseason, very large)
Hummi Grande (midseason, very large)
Gorella (late, very large)
President Carnot (early)

* Not included in main variety list. For those who like to experiment, the English varieties (noted) are available from Sutton & Sons Ltd., Reading RG6 LAB, England; the remaining French varieties can be had from Vilmorin, 4 Quai de la Mégisserie 75001 Paris, France. Write for prices, which change from year to year.

*Varieties by Region**

NORTH— Fairfax, Catskill, Surecrop, Dunlap, Suwannee, New Empire, Fletcher, Howard 17, Sparkle, Ozark Beauty (Everbearer)

SOUTH— Suwannee, Pocahontas, Florida 90, Dixieland, Blakemore, Klondike, Missionary, Albritton

EAST— Raritan, Guardian, Jerseybelle, Midland, Vesper, Surecrop, Ozark Beauty (Everbearer)

SOUTHEAST— Pocahontas, Tennessee Beauty, Earlibelle, Ozark Beauty (Everbearer)

SOUTHWEST— Goldsmith, Red Rich, Sierra, Donner, Red Rich (Everbearer)

NORTHWEST— Northwest, Jumbo, Marshall, Burgess, Spring Giant, Red Rich

PLAINS AND MOUNTAIN STATES— Dunlap, Cyclone, Trumpeter, Ogallala (Everbearer)

NORTH CENTRAL STATES— Sparkle, Midland, Surecrop, Superfruiter (Everbearer)

* These are far from the only excellent varieties for various regions; many more are noted in the main variety list. Also, many varieties will do well in regions they weren't developed for—so experiment if a particular berry sounds good to you and isn't recommended for your area.

*Main Variety List: One Hundred Top Berries**

ALBRITTON— a late large-berried variety similar to *Earlibelle*, of which it was a parent; very productive from Maryland south.

ALISO— a midseason variety with medium-sized berries of fair to good quality.

AMERICAN SWEETHEART— medium-sized early berries on a plant well adapted to the Pacific Northwest.

APOLLO— a late variety with large, tart fruits on a productive, disease-resistant plant.

ARAPAHOE (E)— a hardy everbearer with tasty berries; sets fruit on its runners the same season.

ARMORE— a large berry of fair quality produced in late season by a very productive plant.

ATLAS— a midseason variety with medium-sized, fair-quality berries; very productive and disease-resistant.

BADGERGLO— an early, productive variety developed in Wisconsin that bears large berries of good flavor that are excellent for freezing.

BLAKEMORE— an early-ripening productive plant bearing medium-sized, tart berries good for canning, and very drought-resistant.

BRIGHTMORE— a midseason variety with medium-sized berries that does well in the Pacific Northwest.

BRILLIANT— see *Gem.*

BURGESS HYBRID #41— an early variety with good-sized, flavorful fruit; especially good for the Pacific Northwest.

BURGESS SPRING GIANT— similar to the above but with larger fruit.

CATSKILL— very large fruit of good quality borne in midseason on vigorous, productive plants resistant to verticillium wilt; grown on a wide range of soil types from New England to Minnesota.

CHESAPEAKE— a late variety with large, good-quality fruit.

CHIEF BEMIDJE (E)— an everbearer with large, good-quality berries that produces fruit on its runners; developed in Minnesota, where it survives 40° below zero temperatures without protection.

COLUMBIA— a late variety with large fruit of fairly good quality on a hardy plant.

CYCLONE— good-sized, sweet berries of dessert quality on a productive plant that bears early.

DARROW— a very productive early variety, bearing medium-sized, flavorful berries.

DAYBREAK— an early variety with medium-sized, tasty fruit on a fairly productive plant.

DELITE— a late variety with medium-sized, tasty fruit on a plant that bears on its runners even though it isn't an everbearer and is extremely disease-resistant.

DIXIELAND— medium-sized midseason berries of poor quality on productive plants that are very susceptible to verticillium wilt.

DR. BURRILL— see *Dunlap*.

DONNER— a productive midseason variety that does well in California with medium-sized, tasty berries.

DUNLAP (SENATOR DUNLAP, DR. BURRILL, WONDERBEARING) — medium-sized midseason berries with an excellent wild flavor on a very hardy plant that has been a favorite in home gardens since 1890 and is the best yielder under dry conditions.

EARLIBELLE— an early variety with good tart fruit of medium size on a productive plant.

EARLIDAWN— the variety earliest to bear, but berries are very acid and plant is susceptible to verticillium wilt.

EDEN— berries large, tart, and of fair quality; plants vigorous and productive, fruiting in midseason.

EMPIRE— large, good-quality berries on a most vigorous and productive midseason plant.

ERIE— large midseason berries of fair quality on very productive, vigorous plants.

FAIRFAX— next to Suwannee the best-tasting of all berries; an old favorite dating back to 1923 that has dark red sweet fruit which doesn't keep well but is ideal for the home garden; plant is only moderately productive.

FAIRLAND— large midseason berries of good quality on vigorous, productive plants resistant to red stele.

FLETCHER— a midseason variety with medium-sized berries of excellent quality that are good for freezing; well adapted to New York and New England.

FLORIDA 90— a very productive southern variety good for sandy soils; yields medium-sized berries of fair quality that ship well and are much used commercially.

FORT LARAMIE (E)— a hardy, productive everbearer; bears good, medium-sized berries and has survived temperatures of 40° below zero; a heavy feeder.

FORTUNE— an early bearer yielding berries of excellent dessert quality on a fairly productive plant.

GEM (E)— an average-sized everbearer that is on the tart side and doesn't produce much spring fruit; Superfection and Brilliant, incidentally, are open-pollinated seedlings of Gem and are considered to be identical to Gem for all practical purposes.

GENEVA (E)— an everbearer developed by the New York State Experiment Station that produces abundantly, though it doesn't make a thick bed of plants; medium-sized berries with good flavor.

GOLDSMITH— an early variety with medium-sized fruit of good quality that does well in the mountain and Plains states.

GUARDIAN— a very productive midseason variety with large, sweet fruit.

HARVEST STRETCHER— a late variety that produces lots of medium-sized, tasty berries on a disease-resistant plant.

HONEYLUMP (E)— a moderately productive everbearer with very sweet fruit.

HOWARD 17 (Premier)— large, tasty berries on a hardy, productive plant that bears early and is very disease-resistant; grown in eastern home gardens since 1909.

JERSEYBELLE— very large berries of fair quality on a productive plant that bears late and isn't very disease-resistant.

JUMBO— a late variety with giant, tasty berries that does well in the Pacific Northwest.

KLONDIKE— slightly acid berries of fair quality on a productive plant; bears early and is good for the South.

KLONMORE— similar to Klondike but even more adaptable to regions of great heat; fruit is of poor quality.

LATE GIANT— a very late productive variety with large, "wild flavored" berries of good quality.

MARLATE— another very late variety; large, flavorful berries on a disease-resistant plant.

MARSHALL— a midseason variety bearing tasty, medium-sized berries that does well in the Pacific Northwest.

MASTODON (E)— an everbearer with large, tart berries of fair quality.

MIDLAND— very large quality fruit on an early-bearing productive plant that is a sparse producer of runners; fruit freezes well; grown in home gardens from New England to Virginia since 1929.

MIDWAY— a productive plant yielding medium-sized, tart berries of fair quality that freeze well; bears in midseason and is susceptible to leaf spot and verticillium wilt.

MISSIONARY— a productive midseason variety with medium-sized berries of fair quality; does well along the Gulf Coast.

MORTGAGE LIFTER— very large, tasty berries on vigorous, productive plants that bear in midseason.

NEW EMPIRE— an improvement of Empire by the New York State Experiment Station; bears fruit over a longer period than any variety except everbearers.

NORTHWEST— a midseason variety with medium-sized, tasty berries on a plant well suited to the Pacific Northwest.

OGALLALA (E)— a particularly tasty everbearer which contains the rich, tangy flavor and hardiness of the Rocky Mountain strawberry with the large size and productiveness of the best cultivated varieties; fruits on runners and bears fruit sixty days from planting.

OZARK BEAUTY (E)— the most productive of all everbearers and probably the best all-around everbearer; an easy-to-grow plant that is very disease-resistant, doesn't fruit on runners, and yields large sweet berries of good quality.

PAYMASTER— a heavy-yielding midseason variety with large berries of good, sweet flavor.

POCAHONTAS— large, tart berries of good quality on a midseason plant that is very productive and good for the South but susceptible to verticillium wilt.

PREMIER— see *Howard 17*.

PUGET SOUND— a midseason variety with medium-sized fruit of good quality on a plant well suited to the Pacific Northwest.

PROGRESSIVE (E)— an old everbearer good for the North; yields medium-sized, slightly acid berries of fair quality.

QUINAULT (E)— a tasty everbearer that does very well in California; large, good-quality fruit on a plant that is very disease-resistant and productive, fruiting on its runners.

RARITAN— a productive midseason variety with large, tasty berries.

REDCHIEF— another productive, disease-resistant, midseason type yielding large, good-quality berries.

REDCOAT— a midseason-to-late variety developed in Canada; large berries of good quality on a very hardy, vigorous plant.

REDGLOW— large, good-quality berries on a vigorous, moderately productive plant that bears early.

RED RICH (E)— a prolific everbearer yielding medium-sized, tasty fruit.

REDSTAR— probably the best of the hardy, very late varieties; large, tasty berries on a productive plant; tolerant to virus diseases and leaf spot.

ROBINSON— very large berries of fair to poor quality on a productive plant that likes light soils, bears in midseason, and is very drought-resistant.

ROCKHILL (E)— an oldtime everbearer with berries similar to Progressive, the plant not as hardy.

SENATOR DUNLAP— see *Dunlap*.

SEQUOIA— a new everbearer strictly for the home garden because it doesn't ship well; medium-sized, very sweet, delicious berries on a fairly productive plant.

SNOW KING— a midseason, moderately productive variety that bears *white* fruit of good quality.

SPARKLE— medium-to-large berries of good quality on a vigorous, productive plant that bears late and has long been a great favorite of gardeners.

STELEMASTER— similar to Blakemore but immune to red stele disease.

STOPLIGHT— a very productive early type with large, tasty berries on a disease-resistant plant.

STREAMLINER (E)— a productive everbearer yielding large, flavorful berries.

SUGARBALL (E)— a prolific everbearer that bears large, sweet berries on its runners.

SUNBURST (E)— a new everbearer developed by Burgess with very large, flavorful berries on a hardy, productive plant.

SUPERFECTION (E)— a heavy-bearing everbearer, especially in the fall; yields large, flavorful fruit on a very adaptable plant (see *Gem*).

SURECROP— large, good-quality berries on a moderately productive plant that bears in midseason; immune to red stele disease and resistant to verticillium wilt, leaf spot, and drought.

SUWANNEE— probably the best-tasting berry of all; a midseason type that yields good-sized berries and has been grown in the North and the South since the turn of the century; Professor George Slate, a connoisseur of good berries, has called this "the best-flavored of all strawberries," but others pick Fairfax.

TEMPLE— a midseason variety similar to Sparkle.

TENNESSEE BEAUTY— large, tasty berries on a midseason variety that is very productive but needs a lot of moisture.

TITAN— an early disease-resistant type with very large, good-quality berries.

TORREY— a midseason variety good for the South; medium-sized fruit on a productive plant.

VESPER— a very late productive variety yielding medium-sized fruit of fair quality; isn't very disease-resistant.

WILLIAM BELT (BELT)— large berries of good quality on a late-fruiting variety that doesn't like drought.

WONDERBEARING— see *Dunlap*.

* (E) Indicates an everbearer.

When, Where, And How To Plant

Strawberries are best planted in early spring—as early as the ground can be worked in all sections of the country except the South, the Southwest, and coastal areas of California, where they should be planted in the fall. In the South, especially in Florida and along the Gulf Coast, they can even be planted in early winter.

Choosing a Planting Place: Sites and Soil

In selecting a site for your plants, be sure *not* to choose an area where tomatoes, potatoes, peppers, eggplants, or any member of the Solanaceae family has been grown within three years—for strawberries can contact the soil-borne disease verticillium wilt left by these crops and become stunted or die. It isn't really wise to plant strawberries where *any* crop has grown for the last three years, as such land often contains troublesome pests (especially perennial weeds, insects, and nematodes) as well as soil-borne diseases. If you plan to replace the lawn with a strawberrry bed, turn the sod over at least a year beforehand to rid the area of grubs that may have been feeding at the grass roots and would harm the plants. Where grubs are a minor problem, just turning the soil over and allowing winter cold to kill most of them will be enough. In really infested areas chemicals like chlordane will have to be used.

Strawberries should ideally be grown on newly planted land, but since this is usually impractical, try to pick any plot with a good supply of organic matter that is relatively free of weeds. Definitely avoid soils infested with perennial weeds like nutgrass, quackgrass, or bindweed unless you can entirely eliminate these pests. Strawberries need full sun to do their best, but in selecting the site also try to consider air and water drainage, land slope, and direction of land exposure. Where late spring frosts are frequent, choose ground higher than the surrounding areas, there being less danger of frost damage on high ground because cold air flows to the adjoining low ground. While bottom lands can produce higher yields because of greater soil fertility and moisture, the frost hazard is greater in such locations, too, a slope of only two or three feet in a hundred giving some protection from frost. A gradual slope is preferable to a steep one because it is less liable to soil runoff. Strawberries on southern slopes ripen a few days earlier than those on northern slopes, so choose a site that slopes to the south if you want berries as early as possible and choose one that slopes to the north if you want to delay ripening several days.

Any good garden soil or soil containing an ample supply of organic matter will do for strawberries. Ideally the soil's pH should range from 5.5 to 6.5, moderately acid, but the berries aren't too particular about this. Extremely light and heavy soils aren't as desirable as sandy, gravelly, or silty clay loams. Growers desiring early fruit generally prefer sandy soils, although they are usually more infertile and subject to drought. It is important that strawberry soil be well drained (moist but not wet), as plants can die when wet ground freezes in the winter, especially if the soil is clay or fine sandy silt. Wet soil also inhibits plant growth and can lead to damage by red stele root rot.

To be sure that the soil is fertile, work in compost or cow manure at a rate of one bushel for every 50 square feet. Chicken manure or hog manure should be used at only one-third this rate, as it contains too much available nitrogen. Another way to make certain the soil is well supplied with humus is to plant a green cover crop like clover, rye, or vetch the previous year and turn it under before setting out your strawberries.

If manure isn't available, other organic materials can be dug into the soil. Sawdust, wood chips, or crushed corncobs can be applied at the rate of 10 pounds per square yard. Mix in about 10 ounces of fish meal or 14 ounces of cottonseed meal to provide nitrogen to help break them down. Shredded, composted leaves need no nitrogen added and are applied at the rate of 6 bushels per 100 square feet.

If you want to use commercial fertilizer, which I don't recommend for long-term soil fertility, fertilize with about 2 pounds of 10-10-10 per 100 square feet well before planting. Generally, the more peat moss, compost, and well-rotted manure you can work into a strawberry patch at the onset, the bigger and better your crop from year to year.

Spacing the Plants

Before doing any actual planting, decide which strawberry spacing system you want to use. The four most popular follow, each with its own particular assets and liabilities. Remember that in all these systems, plants are set out in spring and flowers are removed from the plants the entire first growing season so that they can build up strength to yield a bumper crop the next year. This is the case with all varieties except everbearers, whose flowers are removed up until the latter part of July, the plants allowed to blossom and fruit thereafter. Strawberry plants set in the fall can be allowed to fruit the next spring, but yields won't be as high as when they, too, are deblossomed. There are, of course, numerous variations you can make on any of the following spacing methods, many of which often go by other names.

THE HILL SYSTEM. This method requires the most work and yields the biggest berries. Under the hill system plants are set out 12 inches apart in rows 12 inches to 2 feet apart. No new runner or daughter plants (long, green, stringlike growths which produce new plants at their tips) are allowed to form from the mother plants; any runners are snipped off or cut off with a hoe before they can root. Runners can also be prevented from rooting by planting the mother plants in holes made in black plastic spread over the strawberry patch.

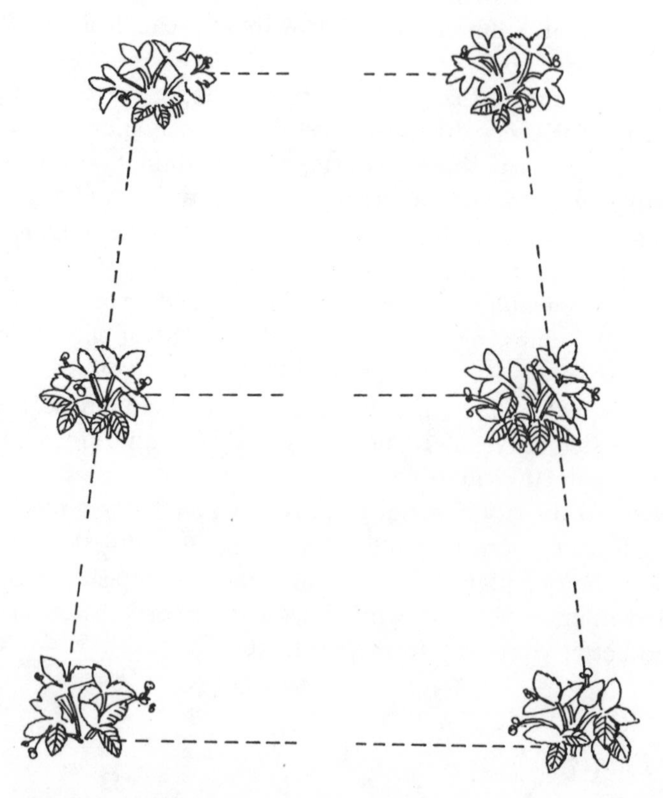

Hill System. All plants are set 12 inches apart and no runners are allowed to develop.

In what is called the *double-row hill system* the plants are set out in two rows 12 inches apart from each other, with the plants 12 inches apart in each row. A 24-inch alley where nothing is planted separates each double row from the next double row. In the *triple-row hill system* three rows of plants are set 12 inches apart from each other, again with the plants 12 inches apart in rows. There is also a 24-inch "alley" separation here between each set of triple rows. A *four-row hill system* can also be used. Obviously there is more bending and stretching the more rows you plant in the hill system, which somewhat offsets the space and productiveness gained.

Plants grown in any hill system are turned under after 2 to 3 years, when the plants lose their productiveness—the first crop will be best and later crops successively poorer. Some growers even replant every year, either from new plants or plants raised from runners. The *triple-row hill system* referred to is used by those California growers who produce 50,000 quarts of berries an acre. The California plants are everbearers that do not set fruit on runners the same season and they are grown in raised beds, which you can simulate by enclosing your patch with railroad ties, filling in good strawberry soil between them, and then setting your plants in the *triple-row system*.

Some growers try to renovate hill system patches after the big-bearing year by mowing and fertilizing the plants immediately after harvest. In any case, the following year's yield is never even half of the prime crop.

THE MATTED-ROW SYSTEM. Here, in the exact opposite of the hill system, the mother plants are set out 2 feet apart and all runner plants are al-

Double-row Hill. Here the plants are set out in 2 rows each 12 inches apart; between double rows there is a 2-foot alley for walking.

lowed to root, making for small berries. Use varieties that form many runners. Everbearers that fruit on their runners the same year should always be grown by this method or the following one. The beds formed usually have 2 to 3 inches between plants and only enough plants are cut off to allow 1½ to 2 feet between rows, each row kept about 2 feet wide. This is the easiest way to grow strawberries, but yields are not great and the quality of the fruit is not excellent. The yield is only greater than the *single-row hill system,* all others topping it. New beds are started anywhere from 2 to 4 years after planting.

THE SPACED MATTED-ROW SYSTEM (or HEDGE-ROW SYSTEM). This is a compromise between the first two methods. Plants are set as in the matted-row system above and 4 to 6 runner plants are allowed to root from each mother plant. The runner plants form a circle around each plant and are 6 to 8 inches apart from each other. The next year these runners will produce a bumper crop and send out more runners. Let all of them root except any that extend into the alleys between rows. Allow the plants to grow for another year or two as if they were planted in a matted row and then turn the bed under, starting a new patch in a different place. It's a lot of work keeping the runners thinned so precisely the first year, but in the second year this system will yield more and better berries than the matted-row system or hill system. Berries will be bigger the second year than in any method but the hill system. After the second year you'll obviously get the same kind of crop as from the matted-row system. Earlidawn, Red Rich, and other varieties that form few runners should be used here.

THE THREE-ROW BED SYSTEM. Rows of plants are set one foot apart, with plants three feet apart in each row. Two of the earliest runners to appear on each mother plant are allowed to root—one to the left of the mother plant down the row and the other to the right of the mother plant up the row. (These runners can be set in place with a hairpin to keep them straight in line while they root.) When they root, they are severed from the mother plant, which is destroyed the second year, the runners replacing it. In following years enough runners from plants are allowed to root to maintain a bed with plants set a foot apart, eliminating all need of replanting.

Care Before Planting

Strawberry plants are usually sold bareroot in bundles of 12 to 100 and sent to you at the time you indicate. Each plant should have a good, healthy, vigorous mass of roots, not a few thick, straggly roots. Soak the plants in water or "puddle" them in a mixture of mud and water for several hours if they arrive in a very dry condition, storing them in the shade.

If you expect 2 to 3 days to pass between receiving plants from the nursery and planting, they can be stored in the refrigerator unopened. But if any longer time is involved, it is best to heel them into the ground individually by placing the plants in a V-shaped trench deep enough for the roots to

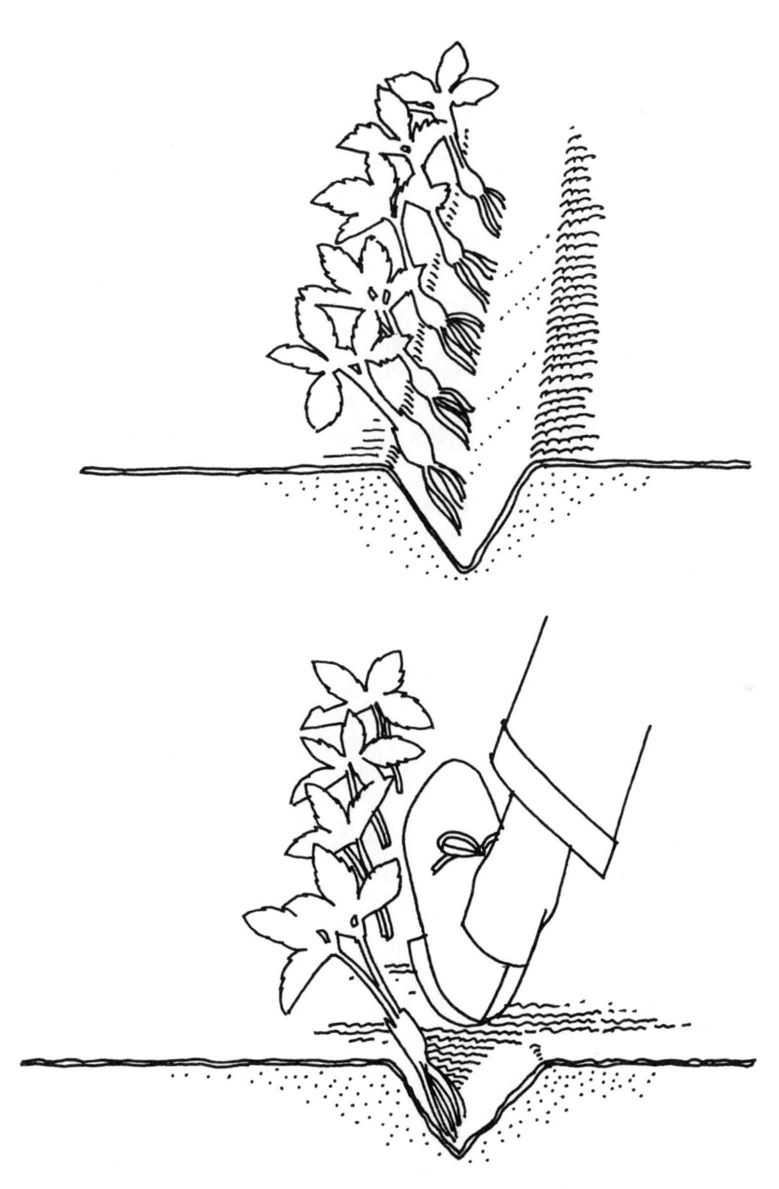

Heeling Strawberry Plants in a V-shaped Trench. To heel in plants, place them in a trench with a crown at ground level (top) and firmly pack soil about the roots (bottom).

be spread out when the crowns are at ground level. Place the plants one next to another along a sloping edge of the trench, taking care not to let the roots of adjacent plants get tangled. Then pack soil firmly about the roots. New roots may form if the plants are heeled in more than several days. Take special care not to damage these when removing the plants from the trench for permanent planting.

How to Plant

Dip roots of the plants in water or a mixture of water and mud while planting and keep them covered so that they aren't exposed to air or sunlight. Space plants apart from one another according to the spacing system you have chosen for your strawberry bed. When planting, remove all but one or two small bright green inside leaves from each plant, trim dead or broken roots off each plant, and prune the roots so that they are no more than 5 or 6 inches long. Make a slit in the soil by inserting a broad-bladed garden trowel about 6 inches deep and moving the handle back and forth. Planting depth is very important—the plants must be set at the same depth they grew in the nursery, not too deep (causing them to smother and die), or too shallow (causing them to dry out). The crown or thick portion in the center of each plant must be just below ground level, half of it buried and half above the soil: *the roots shouldn't be seen.* Keeping this in mind,

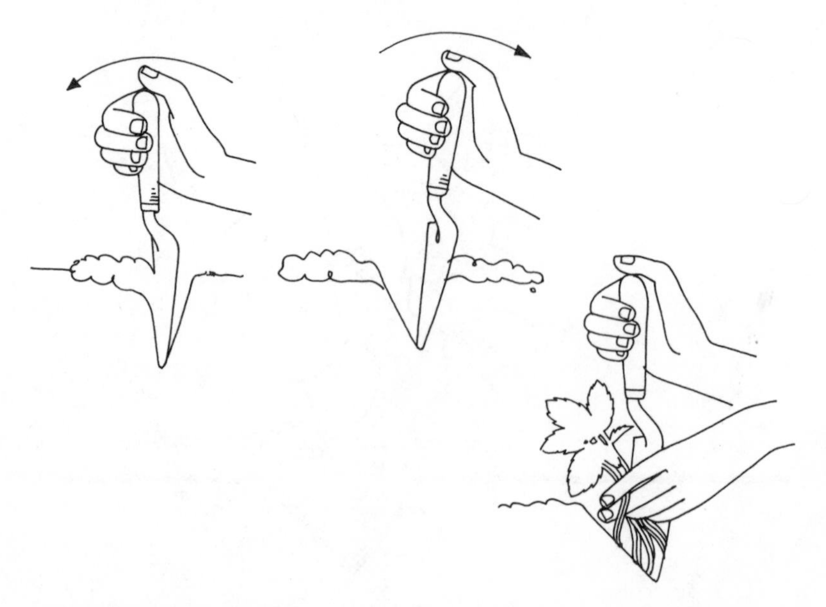

Planting Strawberries. Make a hole with a trowel. Firm each of the sides. Spread roots with your fingers as you place the plant in the hole; fill with soil.

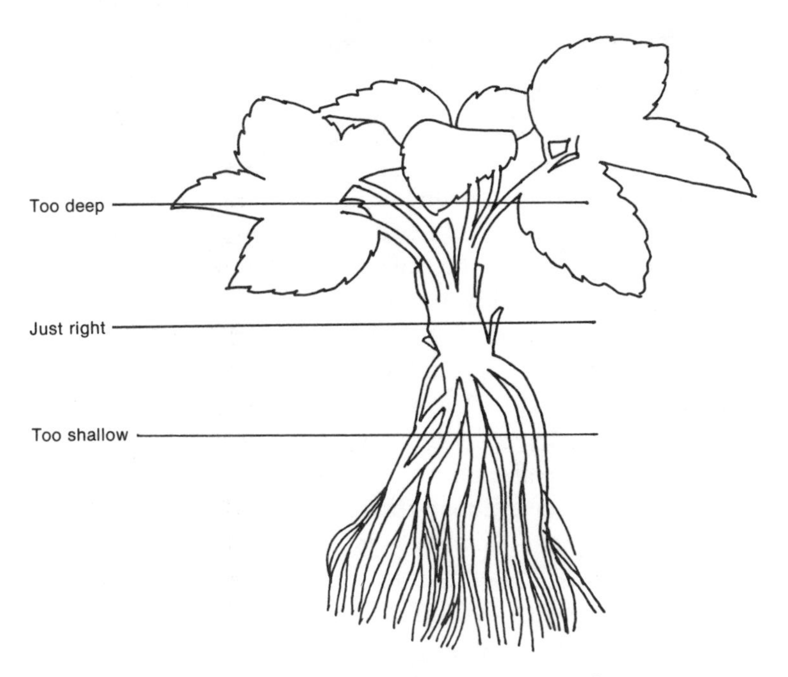

The proper planting depth for a strawberry plant.

set the plants in the slits or holes you have made and spread out their roots in fanlike fashion. Be sure to pack the soil around the roots of each plant to avoid air pockets. One way this can be done is by stepping on each plant—place your instep over the crown of the plant and step firmly. Any plant that can easily be pulled up by a quick jerk on a leaf hasn't been set firmly enough—the leaf stem should break.

Two people can set a large strawberry planting much easier than one, and with much less bending. One person inserts a spade in the soil and the other places a plant in the hole made. The first person then withdraws the spade, firming the soil around the plant roots with his foot.

Water your strawberry plants thoroughly after planting if the soil is dry. The bed won't look like much that day, but within a week the first set of three new leaves will begin emerging from the crown on each plant.

Care Of Strawberries

Mulching Strawberry Beds

Mulching is strongly recommended for strawberries, as it helps keep down weeds, conserves moisture, feeds the plants, and keeps berries clean. It also protects plants against low temperatures and soil heaving. The plants should be mulched right after planting—at least before really hot weather is expected—unless black plastic film is used, in which case the plastic is spread over the patch, anchored down with soil or stones, and the plants are set in the ground through holes cut in the plastic. Good mulching materials include clean straw, salt-marsh hay, pine needles, very strawy manures, leaves, cottonseed hulls, peanut shells, bagasse (sugar-cane fiber), grass clippings, and even old newspapers and rags (a more complete list can be found in Appendix II). Apply the mulch to a depth of 3 to 4 inches after the ground has been thoroughly watered, covering the soil all around the plants, but *not* the plants themselves. About the only trouble you'll have all season will be with runners sent out by the mother plant. These new plants will find it hard rooting in a mulch, and if you want your plants to multiply you'll have to help the runners by moving the mulch aside a bit so that they make contact with the soil and root. By the approach of winter most mulching materials will have about rotted out. At this time, after the temperature reaches 20° F., completely cover the plants, as described in Winter Protection (see Index). This winter covering can be used as next year's mulch when it is pulled off in the spring.

Weeding

Nothing is more important in growing strawberries than keeping weeds down in the bed. This can be done by mulching (see above), but most growers don't practice mulching the first year and many never mulch their plants. If you don't mulch the first year or any year thereafter, constantly keep after the weeds in the patch. Hoe, rake, or weed by hand as often as is needed—at least once every two weeks—to clean out all weeds between plants—otherwise the planting will become a disease-ridden, inextricable

To cut down on weeds strawberry plants are planted in clear plastic and watered with a soaker hose. Plastic is later covered with a bark mulch to prevent overheating in the sun.

mat that is almost impossible to cultivate and bears very little quality fruit. Some gardeners use Sesone, a pre-emergence weed killer, to destroy germinating seeds and small seedlings (it won't destroy established weeds), applying it with a watering can in early spring and late August. Other weeds are killed by applications of CIPC. Organic gardeners obviously will have nothing to do with such chemicals and depend on mulches or their backs. Organic gardeners with large plantings sometimes employ geese to do their weeding. Five to eight young goslings an acre are used to remove grasses (they won't eat most broadleaf weeds), but they must be evicted from the planting during the bloom and harvest seasons or they will eat the flowers and fruit.

If you cultivate strawberry plants with a hoe or tined fork, be careful not to cultivate too deep. Strawberries are shallow-rooted plants, with 95 per cent of their roots in the top 9 inches of soil and few roots penetrating deeper than 12 inches. Hoe toward the plants when cultivating, to keep the roots from being killed by exposure to air, and keep the crowns of the plants at ground level at all times.

It goes without saying that weeding will be much less of a problem if a proper site for the strawberry patch is chosen—one with few weeds—and if any weeds present are eliminated before planting (see Choosing a Planting Site, in the Index).

Thinning Plants and Removing Blossoms

Strawberry plants and runners should be thinned according to the spacing system used (see Strawberry Spacing Systems, in the Index). During the first growing season, all flowering stems on the plants should also be removed so that no berries can form. Steel yourself to do this, even though it hurts, for it will greatly strengthen the mother plants and increase the number of daughter or runner plants, which bear the most fruit the following year. In other words, by pinching off the first year's blossoms, you are assuring a large crop next year instead of two meager crops. Entirely remove the flower *stems* as soon as they appear, preferably before the blossoms atop them open. Do this by pinching the bottom of the stem between your fingernails or snipping them with shears—never try to pull the stem off or the plant might be pulled out of the ground.

Everbearing strawberries are the only exception to this rule. Flowers should be removed from them the first year until the end of July. After that let fruit form to be picked in the fall.

Fertilizing and Watering

There are two schools of opinion about fertilizing strawberries. One holds that if the strawberry patch is prepared properly and the plants are growing vigorously with dark green leaves, they have enough fertilizer. Others feed each plant with about a pint of half-strength balanced liquid fertilizer immediately after planting, and fertilize a second time in late summer by scattering 5-10-5 or 10-10-10 fertilizer around the plant at the rate of one pound per 100 square feet, working it into the soil or spreading it atop the mulch and watering it in. Still other gardeners fertilize every two weeks until the end of August with a balanced liquid fertilizer or liquid manure.

After the first year, strawberries should never be fertilized in the spring prior to fruiting—this leads to excessive green growth at the expense of fruit set. If a patch is renewed after a year or so, it should be fertilized as above.

Strawberry plants usually require more water than they get from the average rainfall, although some varieties like Dunlap and Robinson do very well even when drought strikes. Commercial growers generally irrigate once every 3 to 4 days. Take a tip from them and see that the plants are watered deeply once a week if possible, especially if you don't mulch. An inch of water a week will produce fine, large berries.

Winter Protection

Strawberry plants don't have to be protected in the South, where winter temperatures are normally not low enough to hurt them, but in the North they should be covered as soon as a temperature of 20° F. has occurred and the plants have hardened off. This is about November 16–25 in the New York metropolitan area, earlier as you go farther north. Some varieties (see the main variety list) do withstand temperatures of 40° below zero, but covering plants in winter not only protects their sensitive crowns, it prevents alternate freezing and thawing of the soil that heaves plants out of the ground or breaks their roots, and often kills them. Cover the plants with about a four-inch blanket of either clean straw, salt-marsh hay, pine needles, or leaves (which aren't the best choice, for they tend to mat and make it hard for the plants to push through in spring). Leave the covering on until the plants begin to grow in early spring, when it should be removed so that the plants aren't smothered. The covering can be pulled into the alleys between rows or stored for use again the following winter.

Spring Frost Protection

Strawberry blossoms are killed at temperatures below 31° F. and unexpected frost can ruin an entire spring crop. If a spring frost is predicted after you have removed the winter covering from your strawberries, cover the plants with it again. Or use plastic film held down by rocks or boards, old blankets, burlap bags, old sheets—whatever you have handy. Another method is to spray your plants with a fine spray of water from the garden sprinkler, which will often keep frost off the blossoms by raising the air temperature around the plants. Leave the sprinkler on until the temperature rises above freezing and any ice on the plants disappears.

Eliminating Insect Pests And Diseases

Follow these general sanitary practices and you'll have little trouble with insect pests and diseases in the strawberry patch:

- Don't plant strawberries in an area where any crop has grown for the past 2 to 3 years if possible; definitely don't plant them where members of the tomato family have grown for 2 to 3 years (see Choosing a Planting Site, in the Index).
- Choose a weed-free site or rid the area of all weeds.
- If planting on a lawn, turn the lawn over at least a year beforehand to rid the area of any grubs possibly feeding at grass roots.
- Buy only registered, virus-free (VF) plants, which are widely available from nurseries. Such plants, introduced by the USDA, have made serious virus diseases in strawberries a thing of the past.
- Purchase disease-resistant plants whenever possible, and buy varieties recommended for your area, especially if diseases are a problem there.
- Buy from a nursery whose stock is frequently inspected by state plant inspectors and is certified apparently free of diseases and insect-pest infestation.
- If plants you purchase appear to be infected in any way, don't plant them—send them back to the nursery, or write a letter of complaint asking the nursery to rectify the situation.
- Keep the garden clean and destroy all plants that become infected with incurable diseases.
- Move the strawberry patch to a new location every 3 to 4 years to discourage insect pests or disease buildup in the soil.

Despite all the precautions above, you may have some trouble with insect pests and diseases. The two following lists give descriptions of common strawberry pests and diseases, including symptoms and suggesting various remedies, where there are any:

A List of Strawberry Insect or Animal Pests and Controls

Birds. Leave ripe berries unattended and birds will make a meal of the patch before you can pick it. There are numerous ways to protect against them. Often a transistor radio left playing in the garden will scare birds off. Records of bird distress signals, timed fireworks, and various commercial noisemakers have proved successful in some cases. White string wrapped around plants apparently resembles spiderwebs to birds; they mistake pieces of rope or garden hose for snakes; and broken mirrors, aluminum pie plates, or streamers of cloth or aluminum foil hung in the garden sometimes scare them off, too. Yet these controls, like the time-honored scarecrow, soon become so familiar to the feathered felons that they become contemptuous of them. Volch oil sprayed on berries does keep birds away, but the very best control is to cover the strawberry bed with cheesecloth, clear plastic, or even old nylon curtains when the berries begin to ripen. Be sure to weight the covering down with boards, bricks, or rocks. The wind can blow it away otherwise and I have seen birds actually pull unweighted cheesecloth aside to get at berries. (See Appendix I for wild berries you can plant to divert birds from the crop.)

CROWN BORER. Adults are small, brownish, red-snouted beetles that feed on foliage and berries. Larvae are little white legless grubs, curved and plump, that tunnel through the crowns, cutting off plants. The chemical control is to dust with chlordane around the crowns before the blooming period. Organic control is to destroy the infected bed and start a new one at least 300 feet from it.

CUTWORM. Fleshy green to black striped worms that cut young plants at ground level. Chemical control is usually dusting the ground with chlordane before planting. Organic controls are many and include: 1) placing cardboard collars around plants so that cutworms can't reach them; 2) scattering mothballs or blood meal around to repel cutworms; 3) digging up the ground in early spring to expose and kill cutworms; 4) lining boards over dampened soil to lure cutworms underneath them where you can destroy them; 5) spraying with the biological insecticide Bacillus thuringiensis.

CYCLAMEN MITES. Barely visible white, green, or brown mites which feed at the base of plants on leaves and flowers, destroying both. Newly expanding leaves may turn brown and die. Older foliage turns dark green with a puckered, twisted appearance and plants can be stunted, producing few runners and less fruit. Serious infestations of these mites in the first year often result from infested stock bought from nurseries—all the more reason to deal with a trusted supplier, for the grower can't tell whether or not there are eggs in the unopened leaves and plant crown when buying the plants. Chemical control is to spray with malathion, but this may kill the predatory insects that usually keep mites in control. Try spraying with safe rotenone, a commercial preparation made from tropical plants. Spraying plants forcibly with water is a control worth a try, but try to hit the undersides of the leaves.

LEAF APHIDS. Small greenish plant lice that travel from plant to plant and transmit virus diseases. Like the root aphid they are transported by ants. Use the controls for the Strawberry Root Aphid, below.

LEAF TIERS. Large, pale greenish-yellow worms striped lengthwise with green and white. Curl up leaves, which they feed on and web together. Try dusting with safe pyrethrum twice in a half hour when first noticed.

NEMATODES. Nematodes are microscopic "eel worms" that live in the soil. The *root knot nematodes* form swellings or galls in strawberry plant roots, weakening and stunting the plants, which produce few runners and have lower yields. *Root lesion nematodes* enter and feed on strawberry roots, leaving wounds where soil fungi may enter and cause extensive rotting, making the plants less productive and sometimes killing them. Both of these nematodes cause more severe damage in light, sandy soils. The only chemical control is to fumigate the soil, an expensive, dangerous operation. Organic ways are plentiful, effective, and much safer. They include:

- Rotating crops so that nematodes don't have a chance to build up in the soil.
- Keeping the soil well supplied with humus by digging in organic matter.
- Using a repellent mulch (grass, decayed leaves, or water hyacinths) that encourages the growth of fungi that attacks the nematodes.
- Planting African marigolds near the strawberries. These exude a substance from their roots that will repel nematodes a year after planting.
- Placing slices of the wild or mock cucumber (*Echinocystis lobata*) around the plants to repel nematodes, or soaking the soil with water that asparagus has been cooked in. The USDA has even found that 5 pounds of sugar mixed with 100 pounds of soil will kill all nematodes within 24 hours.

RABBITS. Eat leaves of young plants. Sprinkle blood meal mixed with a gallon of water around the garden, or use Epsom salt sprays, or mothballs. A bucket of hot water in which a chunk of liver has been soaked for thirty minutes is also said to make a good rabbit-repellent spray. Some gardeners bury bottles filled with water around the garden, just the bottle necks sticking out of the ground—the lights reflected off the glass or the wind whistling over the bottles scares the rabbits.

RED SPIDER MITES. Minute reddish-green mites that suck juices from leaves, causing them to brown or yellow, and spin a thin web over leaves. Plants become generally unhealthy. Chemical control is spraying with malathion. Try a jet of plain water aimed at the mites, or spray with safe rotenone.

ROOTWORM. Small, chunky, copper-colored beetles which feed on foliage, their white, brown-spotted grubs feeding on roots. Both weaken the plants. Chemical control is dusting with chlordane before planting.

ROSE CHAFER. Medium-sized, yellowish, long-legged, clumsy beetles that eat foliage and blossoms. Their whitish grubs attack roots. Chemical control is to dust with a rose dust or all-around dust before buds open. Or try handpicking the beetles.

SLUGS. Large, slimy, soft-bodied, snail-like forms that feed at night. Not a major strawberry problem, but sometimes they will eat foliage and fruit. Slugs hide under rocks, boards, mulch, and other objects in the daytime and travel at night at the rate of a mile in eight nights. There are many, many controls:

- Frequently inspect under mulch and collect slugs, destroying them by dropping them into kerosene.
- Place boards or cabbage leaves in the garden to attract and collect slugs.
- Turn over soil in early spring to expose and kill slug eggs.
- Put down coarse, scratchy mulches like hay, which slugs dislike.
- Use new products like Snail Snare, which dehydrate and kill slugs rather than poisoning them.
- Collect slugs with baits like commercial methaldehyde and destroy them.
- Collect and destroy slugs with shallow pans of stale beer. Slugs will crawl into the pans at night and drown. Other similar homemade solutions include grape wine, blackberry wine, vinegar, and a tablespoon of flour and ⅛ teaspoon of yeast mixed with a cup of water.

SNAILS. Same controls as for Slugs, above.

SPITTLEBUG. Immature forms live in a mass of froth or spittle attached to plants. Adults are grayish, frog-shaped bugs that fly about. Spittlebugs suck the sap from foliage and weaken plants, the fruit on infested plants often small and of poor quality. Control is seldom necessary except when infestation is heavy. Chemical control is dusting with an all-around dust. Try handpicking.

SQUIRRELS. Should squirrels be a nuisance rooting around in the patch for buried nuts or bothering berries, you can sprinkle mothballs or similar commercial preparations around the plants. This will repel squirrels to a degree.

STRAWBERRY LEAFROLLERS. Small green or brownish caterpillars which first occur on the undersides of leaves, then roll or fold the leaves up about themselves. The rolled leaves, fed upon by the insect, turn brown and die, damaging the plants and decreasing the crop. The white web of the leafroller can be seen inside the rolled leaves. Chemical control is spraying with Sevin. Try handpicking or spraying with harmless rotenone.

STRAWBERRY ROOT APHID. Small blue-green, soft-bodied plant lice that first appear on the leaves and are later carried to the roots by ants. They suck out plant juices and weaken plants. Chemical control is to dust soil with chlordane before planting. The organic ways are numerous, including methods designed to eliminate the ants, which keep aphids as "milk cows," transporting them around for the sweet secretion they exude. Organic controls for aphids include:

- Planting strawberries in rich humus soil, which aphids avoid.
- Placing aphid-repelling nasturtiums, garlic, chives, or rhubarb near strawberries.
- Washing the aphids off plants with water or a solution of soapy water.
- Using the biological spray Basic H made from soybeans, or the safe insecticides rotenone and pyrethrum.
- Sprinkling bone meal around the holes of the ants that carry aphids.

STRAWBERRY ROOT WEEVIL. Also known as the "clipper," this is a small, short-snouted beetle, reddish-brown to almost black, that feeds on leaves and berries. Its larvae are white, legless, curved grubs that feed in the crowns. The adult hibernates under the mulch or under plant debris and appears on plants at blossoming time, the female depositing an egg in a bud and then cutting the stem so that the bud falls off or hangs suspended by a few threads of plant tissue. Over half the buds in a bed can be destroyed. Chemical control is chlordane dust applied to the soil before planting. The only good organic control is to plow up the bed after harvest every year and start a new bed more than 300 feet away. This will prevent buildup of the pests.

TARNISHED PLANT BUG. A small brown sucking insect that severely reduces the quality and quantity of strawberry yields, especially of everbearers, which mature in late season when the pest is most active. Injury in berries results from feeding punctures made before the strawberries are ripe. The affected "button fruit" is small and deformed at the tip, the seeds crowded together. (It should be noted, however, that many factors that prevent proper pollination can also cause deformed fruit, even inclement weather.) Chemical control is to spray with malathion or Sevin, or dust with an all-around dust. Also try handpicking the insects.

TWO-SPOTTED SPIDER MITE. A barely visible insect varying in color from pale greenish-yellow to dark crimson, usually with one dark spot on each side of the body. These mites suck juices from the leaves and if uncontrolled seriously retard plant growth and fruit production. Control as with Red Spider Mites, above.

WHITEFLIES. Little white moths that flutter out when disturbed and thus are often called "flying dandruff." Their larvae feed on the undersides of leaves, weakening the plants. Chemical control is spraying with malathion. Effective organic controls are garlic sprays, and planting mint or tansy in the garden to repel the flies.

WHITE GRUBS. Adults are large June beetles. The larvae which do all the damage are big, plump, yellowish-white, curled grubs in the soil that feed heavily on strawberry roots and can seriously damage them. Do not plant strawberries in ground that was sod the previous year—wait at least a year after turning over the sod. If you do plant immediately after sod has been turned over, use a soil insecticide first. Chlordane applied to the soil before planting is the usual choice.

WIREWORMS. Soil-infesting, cylindrical, yellow to brown, hard-shelled, shiny, jointed worms up to 1½ inches long that feed on plant roots. Chemical control is spraying soil with Dieldren 18 before planting. Organic control is not to plant strawberries after soil has been in sod—wait at least a year after turning under the sod. Enriching soil with humus also helps, as does good drainage and planting turnips or radishes as trap crops to collect wireworms. The worms will gravitate to the turnips or radishes and can be easily collected just by pulling up the roots.

A List of Strawberry Diseases and Controls

BLACK SEED. See Leaf Spot, below.

BOTRYTIS FRUIT ROT. Commonly called gray mold or brown rot, this is the most serious and widespread of the half dozen or so rots that attack strawberries. It first appears as a light brown spot on green or ripe berries and progresses through the fruit. Later, the entire berry becomes brown and rotten. During wet periods a gray dusty fungus covers the infected fruit. The fungus spores of botrytis rot are easily detached and carried to berries nearby by air currents and splashing water. Good gardening practices will therefore reduce losses. These include spacing plants adequately; maintaining a narrow matted row; keeping the planting weeded; irrigating only when conditions favor rapid drying; and using mulches that prevent the berries from touching the soil. Also avoid the use of nitrogen fertilizer during the spring before the harvest season. Nitrogen induces dense foliage and tends to soften fruit, making it more susceptible to the disease. Some organic growers claim that the use of dolomitic limestone and colloidal phosphate eliminate botrytis fruit rot. Chemical control is spraying with a fungicide.

BROWN ROT. Same controls as for Botrytis Fruit Rot, above.

DWARFING. Caused by microscopic roundworms which live between the folded young leaves and suck out sap, resulting in dwarfing and wrinkling of new leaves. Occurs in the southern United States. Control is not to plant new beds in old soil and to rogue out all infested plants and destroy them.

GRAY MOLD. Same controls as for Botrytis Fruit Rot, above.

LEAF SCORCH. Irregular, dark purple spots *without* gray centers characterize this fungal disease. Attacks old as well as new growth, and attacks all parts of the plant above ground. Same controls as for Leaf Spot, below.

LEAF SPOT. Only new growth is susceptible. Appears on leaves as purple circular spots with gray centers. The disease also occurs on berry caps, leaf and fruit stalks, and on the berries themselves, where it turns one or several seeds of the berries brownish-black and is called *black seed* or *black berry*. Abundant moisture encourages the development of the disease. Organic protection is essentially the same as for Botrytis Fruit Rot, above. Chemical control is spraying with fungicides. The varieties Catskill, Premier, Midway, and Robinson are resistant to leaf spot. Very susceptible varieties are Jerseybelle, Pocahontas, Redglow, and Sparkle.

POWDERY MILDEW. Leaves curl upward and there is a white cobweblike growth on the lower surface of the leaves. The disease often occurs during extended periods of cool weather in the growing season and is not as damaging as it is unsightly. Control is to plant resistant varieties like Catskill, Empire, and Sparkle. Dixieland, Midland, Redglow, Stelemaster, and Jerseybelle are very susceptible to powdery mildew.

RED STELE DISEASE. A fungus disease that attacks the roots of plants. Plants become stunted, older leaves die, new leaves turn small and bluish, fibrous roots decay, main roots rot, and the central part or "stele" of the root turns dark red. The plants rarely produce a normal crop and often collapse and die during harvest season, just before the fruit ripens. There is no effective control for red stele, which is probably the most serious strawberry disease in America. It goes without saying that growers should avoid planting red-stele-infected stock. There are red-stele-resistant varieties, including Redcrop, Sparkle, Stelemaster, Surecrop, Temple, and Vermillion.

VERTICILLIUM WILT. This common soil-borne fungus disease attacks the leaves and vascular system of the plants, weakening them and decreasing their yield. The earliest symptom is wilting and dying of the older outer leaves. Younger leaves become paler and begin to curl upward. Finally, the entire plant may collapse and die. Chemical fungicides do little good here because this fungus lives so long in the soil. Controls are to plant in ground free of the verticillium fungus. Never plant strawberries on land where tomatoes, peppers, potatoes, eggplants, or other strawberries have been growing for the last two years—for the fungus increases tremendously in their presence. Keep the garden clean and don't dig any of the above plants or their fruits into the soil. Varieties *highly resistant* to verticillium wilt include Catskill, Empire, Erie, Frontenac, Fulton, Premier, Surecrop, and Vermillion. *Moderately resistant* varieties are Fletcher, Blakemore, Guardian, Marshall, Sunrise, Tennessee Beauty, Salinas, and Robinson. *Very susceptible* varieties are Dixieland, Earlidawn, Pocahontas, Armore, Midway, Jerseybelle, Midland, Redstar, Daybreak, Lassen, Northwest, Vesper, and Raritan.

VIRUS DISEASES. A great many virus diseases attack strawberries, some of them producing clear symptoms such as mottling (spotting or blotching) of leaves, and others producing no visible symptoms. Plant vigor is affected, fruit yield is lowered, and plants cannot be cured once infected. Since most strawberry viruses are carried by aphids, they are best controlled by controlling the aphids. But if the gardener buys virus-free (VF) strawberry plants developed by the USDA there is little to worry about concerning virus diseases. These plants are offered by almost every nursery today. If the plants catch a virus in the garden, however, precautions do have to be taken against the aphids.

YELLOWS. A heredity defect, not a virus disease, that is also called *Spring yellows* and *June yellows*. Plants are stunted and yellow. Affected plants usually occur together in spots or rows. Their leaves are rounded, often cupped or twisted, and have yellow margins. The plants never recover and bear less fruit than healthy plants. Certain varieties, such as Dixieland and Earlidawn, are very susceptible.

PROPAGATION AND RENEWAL

Free Strawberry Plants: Easy Propagation Methods

Except for the wild ones (see Index), strawberries do not come true from seed; that is, strawberry seed will yield plants unlike the plant it came from, plants with characteristics of the hybrid plant's ancestors. But it isn't necessary to grow strawberries from seed. As noted, strawberries take care of propagating themselves admirably by sending out runners or daughter plants that root nearby. These can be used to renew the strawberry patch (see Strawberry Spacing Systems, in the Index) by holding them in place with a stone, hairpin, clothespin, or small handful of dirt until they root. Or they can be allowed to root and then be transplanted to a new strawberry bed.

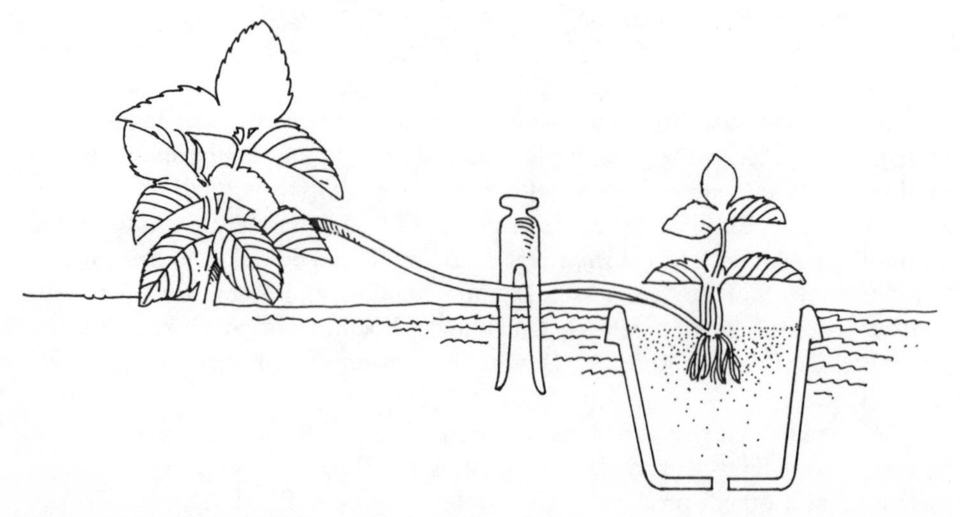

Propagating by Rooting. The strawberry runner plants grow in a pot sunk in the ground, the plants held in place with a clothespin or hairpin.

The best way to do this is to fill peat pots, cut-down milk containers, or small flowerpots with good strawberry soil, sink them in the earth near the runners, and let the runners root one in a pot. Hold each runner in place with a hairpin, clothespin, or stone. When the runners are firmly rooted, sever them from the mother plants and transplant elsewhere. Peat pots can, of course, then be transplanted without lifting the plants from them.

New Life for an Old Strawberry Bed

If you don't want to start a new bed when production begins to fall off in the old strawberry patch, there is an alternative that works at least passably—although you should remember that commercial growers almost always plow under a bed after 2 years and usually do so after the first year's crop is picked. Nevertheless, up to one-half production from an old bed (and sometimes more) can be maintained for 2 to 3 years or longer if the following method is used. Begin the renewal at the end of the harvest season in early summer—don't wait 2 or 3 weeks but get to work as soon as the berries have been picked. At this time run your hand- or powermower, set on high (2 to 3 inches), through the strawberry patch, cutting off the tops

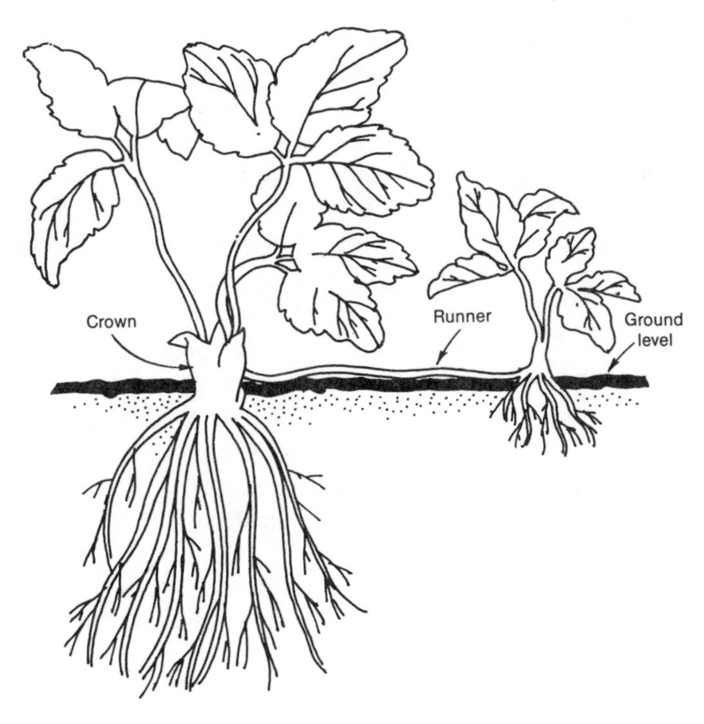

Anatomy of a strawberry plant showing mother plant (left) and runner plant (right).

of the plants (a scythe or hoe will do just as well). The plants will then put all their strength into producing new leaves and fruit buds for the next year (the more new leaves a plant has, the more berries it will produce). Help them along by weeding the patch thoroughly and fertilizing the remaining plants. Also turn under every other row in the patch, including all plants and any mulch that may be present. Runners from the alternate rows will soon fill up these now-empty rows and you will get fruit from both the topped plants and their runners the following season—more from the topped ones. If the plants don't send out many runners, encourage them to do so by digging in a little cottonseed meal around each plant.

This method works best where very productive varieties like Pocahontas have been planted. When you use the hill system or spaced-row system, no thinning of plants is necessary. With the matted-row system, thin plants (that is, pull out excess ones) when weeding until the plants are 6 to 8 inches apart. Sprinkling an inch or so of compost through the bed after renewing is also a good idea. Within 2 to 3 weeks new foliage will appear on the plants, which will look so bad at first that you'll think you made a mistake following my advice, but in another 3 to 4 weeks they will be thriving.

Pickin' 'Em

Strawberries are generally mature about thirty days after blossoming in mild weather, faster when the weather is warm. (Incidentally, the first flower to open on a plant is always the largest flower and always becomes the largest fruit with the most seeds.) Pick strawberries when they are perfectly red and ripe in the home garden, unless you plan to sell them to local markets or from a roadside stand—once they are whitish they will change color and become redder after picking, but will never gain in flavor when off the plant. Fruit to be used for freezing, however, should be firm and not dead ripe, while fruit picked for jelly-making (when concentrated pectin won't be used) should be slightly underripe. Harvest the berries daily or at least every other day, the best time for strawberry picking being in the early morning while the berries are cool. Do not pick on rainy days, as berries picked in the rain may not keep well. Never pull or pick the berries from their stems, either, for you might pull off an unripe cluster in the process. Just pinch the stem between the thumb and forefinger so that the berry is removed with the calyx and a short piece of the stem intact. Place the berries in shallow containers where they won't be bruised piled one on top of another and don't keep them in direct sunlight for more than 10 to 15 minutes.

Strawberry Shortcake And Other Recipes

There are doubtless more delicious recipes for strawberries than for any small fruit. These range from simple strawberries and cream to a reputedly aphrodisiac concoction made of strawberries and pears doused in Cointreau and drenched in a fragrant sauce of beaten egg yoke, confectioners' sugar, cloves, and cinnamon. Strawberries are also a dieter's delight, for one large berry averages only 4 calories and an entire cup amounts to only 55 calories. Then there is strawberry wine, strawberries Romanoff, *fraises à la Cussy*—the simple but elegant strawberries, cream, and champagne dessert invented by Napoleon's chef—and, of course, that American masterpiece, old-fashioned strawberry shortcake. You're sure to find more than one you like from among the following delicacies. Remember never to hull berries or wash them until they are to be used. Leave their caps on and refrigerate them for an hour or so if need be. Then wash them in ice water, drain in a colander, and use a stainless steel knife to hull and slice them if desired. If you remove the caps before washing the berries, they will absorb water.

Strawberry Jam

Said one strawberry to another: "We wouldn't be in this jam now if we hadn't been in the same bed!" For a good strawberry jam place 2 quarts of hulled and crushed berries in a pot and bring to a boil, adding 6 cups of sugar and stirring to dissolve. Next add ⅓ cup of lemon juice. Bring the mixture to a rapid boil until the jellying point is reached, and pour into hot, dry jelly jars. Adjust jar caps and process 10 to 15 minutes in a water bath or top with paraffin wax. Makes 4 pints.

Strawberry Summer Soup

2 pints strawberries
1 cup orange juice
1¼ teaspoons instant tapioca
⅛ teaspoon cinnamon
½ cup sugar
1 teaspoon grated lemon peel
1 tablespoon lemon juice
1 cup buttermilk
2 cantaloupes, chilled
4 thin slices of lemon

Purée all but six berries in blender and strain into saucepan, adding orange juice. Mix tapioca and 4 tablespoons of the puréed strawberry mixture. Add this to the saucepan along with the cinnamon. Stir until mixture comes to a boil, then cook until thickened (about one minute). Pour soup into large bowl and add sugar, lemon peel, lemon juice, and buttermilk, blending well. Slice whole strawberries into soup and chill eight hours. Serve in cantaloupe halves with the seeds scooped out, floating a lemon slice in each. Serves 4.

Frozen Whole Berries

Spread flawless berries on a cookie tray and freeze a tray at a time. Pack them into freezer bags immediately after freezing, seal, and store in the freezer. The frozen berries can also be frozen in a plastic container, but they will freeze more slowly this way and thus be mushier when thawed.

Frozen Sliced Berries

Slice hulled berries lengthwise in thirds, adding one cup sugar for every seven cups of fruit and letting stand 10 minutes until the sugar dissolves. Stir and pack into plastic freezer bags that can be sealed with a heat device and seal. A quart of berries makes a pint and a half of sliced berries.

Frozen Strawberry Purée

A good way to salvage damaged strawberries. Cut blemishes from damaged berries and purée in a blender, adding a teaspoon of sugar and a teaspoon of lemon juice for each pint of strawberries if desired. Pour into glass freezer jars, cover, and freeze.

Strawberry Shortcake

3 pints sliced strawberries
¼ cup honey
4 teaspoons lemon juice
2 cups flour
5 teaspoons baking powder
1 teaspoon salt
3 tablespoons sugar
¼ cup shortening
⅔ cup milk
2 cups whipped cream

Preheat oven to 425°. Combine strawberries, honey, and lemon juice, letting stand for several hours. Sift together flour, baking powder, salt, and sugar. Cut in shortening with a knife. Add milk gradually and mix to a soft dough. Roll out on a floured board to ½ inch thickness. Cut with a large cookie cutter to make 8 shortcakes and bake on cookie sheet for 15 minutes. Break shortcakes apart and put strawberry mixture between the halves and on top of each shortcake. Serve warm with whipped cream.

Chocolate Strawberries

4 dozen good-sized strawberries
2 pounds of semi-sweet Baker's chocolate

Wash and dry strawberries. Chop chocolate in small pieces and melt slowly in a double boiler, stirring constantly, until it reaches 95°. Dip the strawberries into the chocolate mixture and refrigerate on a cookie sheet lined with wax paper.

Cream-filled Strawberries

3 dozen jumbo strawberries
2 cups cream
3 tablespoons cream sherry
¼ cup powdered sugar

Split each berry into quarters, starting from pointed end, but don't cut through the stem. Refrigerate berries. Combine cream and sherry, beating until the cream is partially whipped. Then add powdered sugar and complete whipping. Fill each berry with the mixture.

Strawberry Wine

Good strawberry wine can be made by using the relatively inexpensive home winemaking equipment available today, which comes complete with directions. But here is a century-old recipe (untested here) that may enable you to make strawberry wine without any special equipment:

"Mash the berries and add to each gallon of fruit a half gallon of boiling water. Let it stand twenty-four hours, then strain and add three pounds brown sugar to each gallon juice. Let it stand thirty-six hours, skimming the impurities that rise to the top. Put in a cask, reserving some to add as it escapes from the cask. Fill each morning. Cork and seal tightly, after the fermentation is over."

Strawberries with Liqueur

Strawberries can be hulled, sugared, and served very cold sprinkled with a number of liqueurs, including either kirsch, cherry brandy, raspberry brandy, kummel, cognac, or maraschino. Chilled strawberries sprinkled with champagne is another delicious dish.

Strawberry Ice Cream

1 quart strawberries
1⅓ cups sugar
3 cups heavy cream

Chop strawberries, mix with ½ cup of sugar, and allow to stand for one hour. Scald 2 cups of the cream. Stir in remaining sugar until dissolved. Add remaining cup of cream. Mash the strawberries through a coarse sieve and mix with the cream mixture. Freeze according to directions for your ice cream maker.

Strawberry Ice

3 pints ripe strawberries
3 tablespoons lemon juice
1 cup sugar

Slice berries in thin pieces and let stand at least 30 minutes. Stir in lemon juice and sugar, then use a blender to blend berries and all their juices into a purée. Pour into two empty ice cube trays, cover with foil, and freeze until almost solid. Take out and beat in electric mixer until full and smooth. Beat for the last time 2 hours before serving. Return to freezer in covered containers.

Open Strawberry Pie

1 3-ounce package softened cream cheese
1 8-inch baked pie shell
1 quart strawberries, sliced
1 cup sugar
3 tablespoons cornstarch
1 pint crushed strawberries

Spread cream cheese over baked pie shell. Add quart of strawberries and set aside. Add sugar and cornstarch to pint of strawberries and boil over low heat until thickened. Pour mixture over strawberries in pie shell. Chill and serve with whipped cream.

Easy Strawberry Ice Cream

(Without an ice cream maker)

4 egg yolks
1 cup sugar
2 cups heavy cream
2 cups strawberry juice and pulp

Beat egg yolks with ½ cup sugar. Add remaining sugar to cream in a saucepan and bring to boiling point over a medium flame. Stir cream gradually into egg and sugar mixture. Strain. Cool. Stir in strawberry juice and pulp. Freeze.

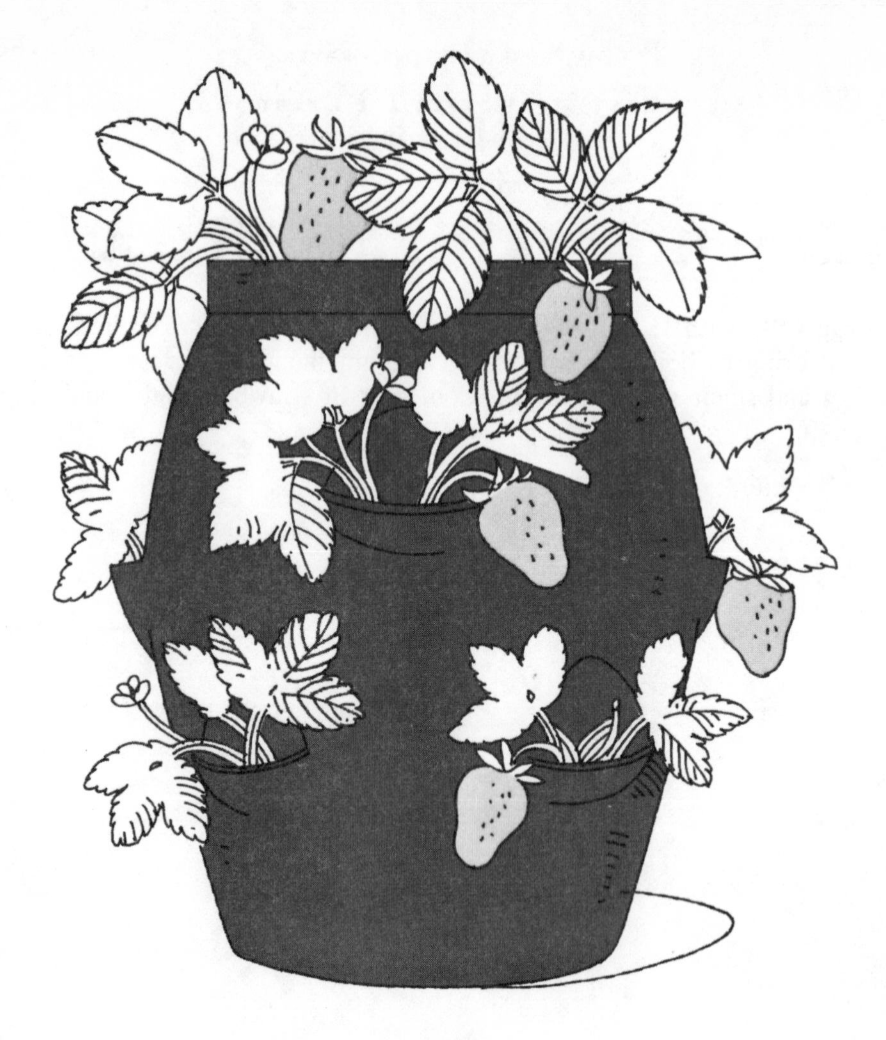

2

Alternate Lifestyles for Strawberries

More than any other fruit the strawberry lends itself well to space-saving planting in the garden, on the backyard patio, and even high up on city terraces. Strawberries thrive in anything from a plain box or jar to a strawberry barrel or large pyramid planter. Recommended varieties for such plantings are the new and prolific everbearers, either Ozark Beauty, Geneva, or Ogallala being good choices. Crops, of course, won't be as large as those from plants grown in a berry patch, but you will be guaranteed some fresh strawberries every summer. The best soil for all planters is a mixture of equal parts of rich loam, manure, compost, and sand. When any of these containers are used, all strawberry plant runners should be removed (although some gardeners root them into peat pots first for use elsewhere) and plants should be replaced every few years. Feed the plants with a complete water-soluble fertilizer. They are easy to protect from birds during fruiting time by covering the container with a cheesecloth netting or punctured plastic sheet.

Space- and Work-Saving Ways To Grow Them

Store-bought Strawberry Pyramids

The strawberry pyramid, or ring as it is sometimes called, allows you to have a strawberry patch on the patio or terrace and makes an attractive display on the front lawn. Prefabricated aluminum pyramids with sprinklers and protective coverings are available from numerous manufacturers. They consist of three rings, each progressively smaller, hold about 50 to 75 plants, and come equipped with their own sprinkler, to which you attach your garden hose. Without accessories like a cheesecloth cover, to protect ripe fruit from birds, they cost from twelve to fifteen dollars.

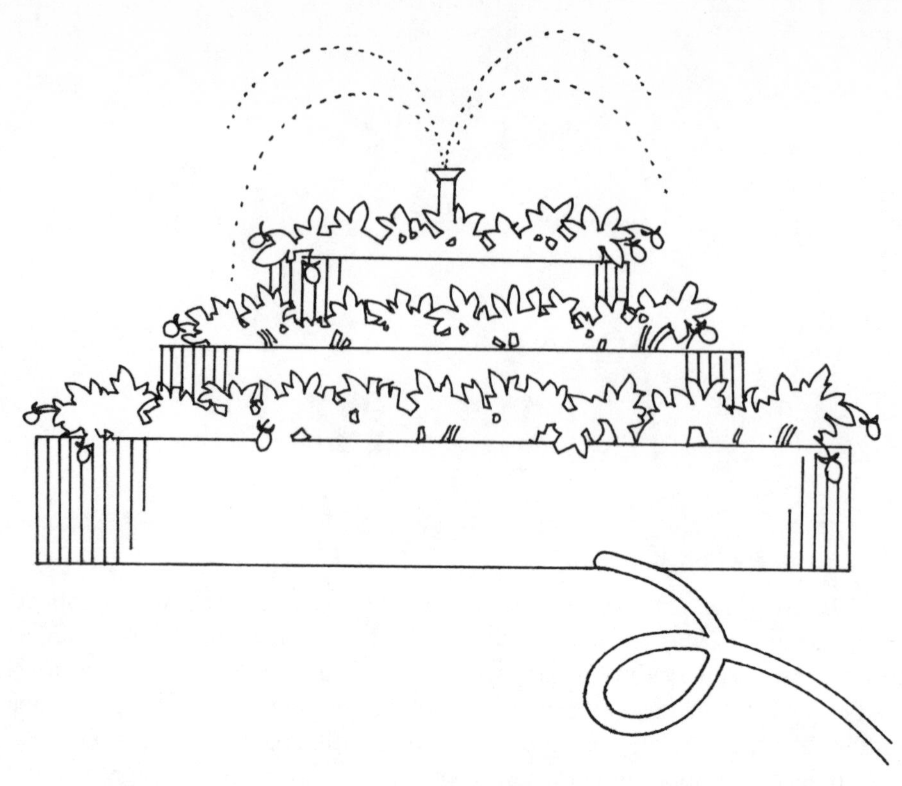

Commercial strawberry pyramid with built-in sprinkler.

Homemade Pyramids

You can make your own strawberry pyramid from either metal or wood. The *circular metal pyramid* is simply a series of three tiers, the bottom one generally 6 feet in diameter, the second 4 feet in diameter, and the third 2 feet in diameter. To construct a strawberry pyramid from aluminum, use corregated lawn-edging sheets, which are the necessary 6 inches or so wide. Build the first tier as a circle 6 feet in diameter, setting it on the ground and filling it with soil. Then add the second and third circular tiers, filling each when it is set in place. A piece of perforated 3-inch pipe can be inserted from the top to water through, or a sprinkler can be placed at the top.

To construct a *square strawberry planter* from lumber, use two-by-eights coated with wood preservative and precut in whatever dimensions you desire—one tier might be 5 feet square, the other 3½ feet square, and the last 2 feet square. Join the boards for each tier frame with ½ by 3½-inch corner braces, using 16d nails. Then place each frame atop the other as with the circular metal pyramid.

Either of these terraced planters will hold about fifty plants. They'll grow even more berries than on the same space in a strawberry patch if you

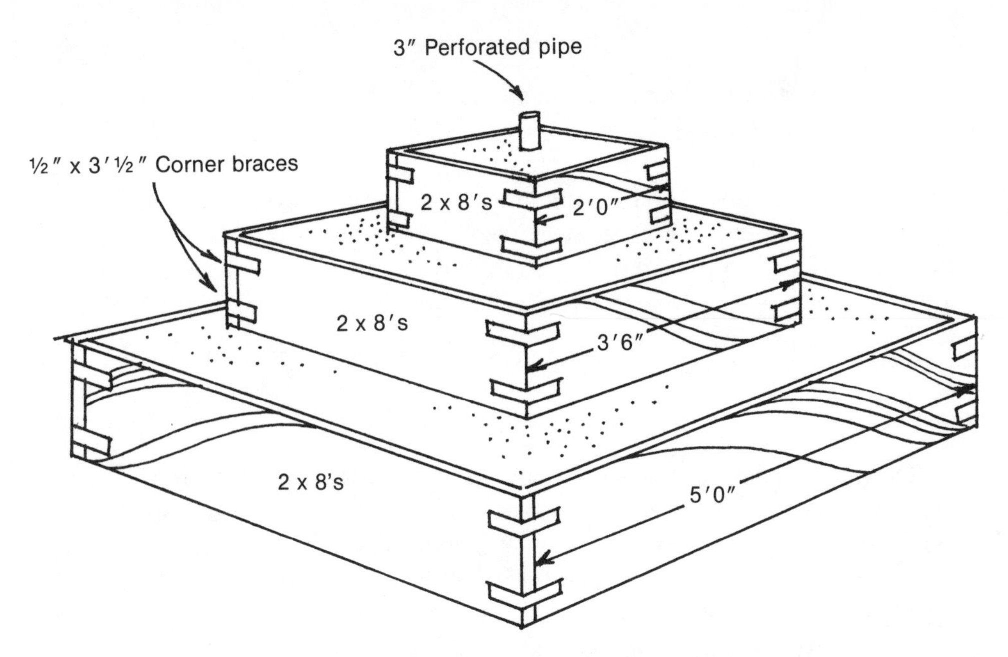

Homemade strawberry pyramid made from lumber.

drill holes in the sides of the wooden or aluminum tiers and plant strawberries therein. Pyramids are all far more convenient to tend than strawberry beds.

Barrels of All Sizes

The old-fashioned strawberry barrel is another great space saver and is portable as well. Berries grown in this way will always be clean and will ripen evenly. Strawberry barrels can be made out of anything from a small wooden nail keg to steel oil drums. Starting at the bottom, simply drill from 2 to 4-inch-diameter holes in circles around the barrel. Stagger each circle of holes 6 to 12 inches from the preceding one. Make the holes in each circle 6 to 12 inches apart from center to center. Finally, drill small holes in the bottom for drainage, and paint or decorate the container for use on the lawn, terrace, or patio. Attach coasters to the bottom so that the barrel can be moved about, or set it on a wagon wheel that spins around, or simply place the barrel on bricks or wooden blocks. Each of these methods provides good air circulation.

When ready to plant, put stones or gravel in the bottom for drainage and fill the barrel with your soil mix. Plant as you fill the barrel with soil, inserting the plants from the inside of the barrel, guiding the leaves and crown outside through the holes and fanning out the roots inside. When the barrel is planted, insert a piece of perforated drainpipe that has been cut to the height of the barrel (rolled screen can also be used). This is filled with pure sand and watered through.

Strawberry barrels can be winter-protected by bringing them into the garage or a cool cellar, or by mulching them heavily with straw or leaves held in place with chicken wire. It is worth a try to save them for next year, but generally the plants should be replaced with new ones after they have fruited, for they bear very sparsely after the first year.

Tubs

These are made in the same way as strawberry barrels, but the container, of course, is smaller and wider. A strawberry tub made of 12 interlocking rigid panels forming a tub 29 inches high and 23 inches in diameter, with room enough for 44 plants, is available from Rotocrop Inc., 604 Aero Park, Doylestown, Pennsylvania 18901, for $19.95. A special soil cover and "easy-watering irrigation tube" are included.

Jars

Probably the oldest of strawberry planters, the strawberry jar comes in several sizes. The clay, terra-cotta-colored containers are offered by some nurseries and are well worth looking for. These attractive ceramic jars have openings or "pockets" in the sides as well as room for a few plants on top. They can be set outside, in a sunny window, or under artificial light and are cared for the same way as strawberry barrels (see Index).

Boxes

If you can't find a strawberry jar, or don't want to buy one, a small box, 2 × 2 feet or longer, will serve just as well for the patio or a sunny window. Either find the right-sized box, or build one to your own specifications. Fill the container with a rich soil and plant a dozen or so everbearers for a refreshing treat all summer long. Place a large stone in a corner of the box that you can run the hose against when watering plants.

A Strawberry Tree

The "tree" is simply half-cylinder wire nailed to a 12-inch board and lined with sphagnum moss to hold your soil mix. The strawberries are

Preparing A Strawberry Jar

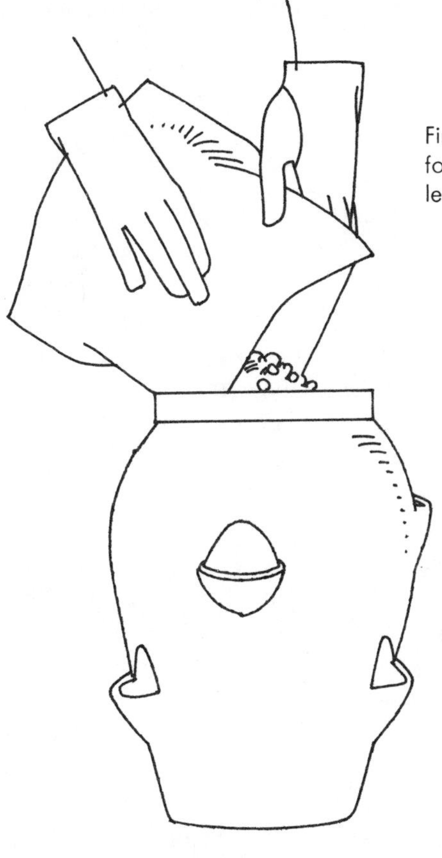

First add a layer of stones or gravel for drainage. Then fill with soil to level of first hole.

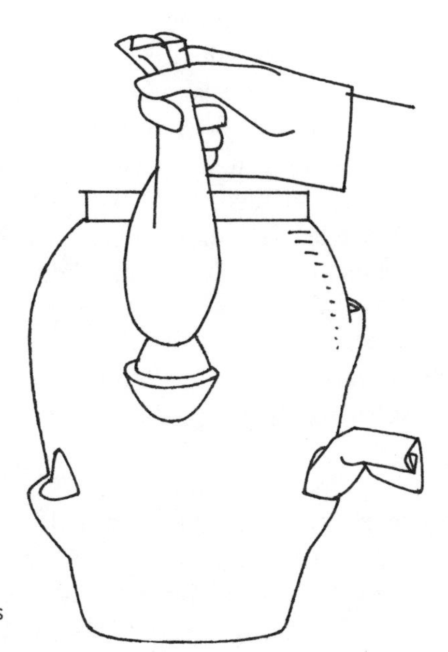

To hold soil in place, fill nylon socks with soil and insert in pockets.

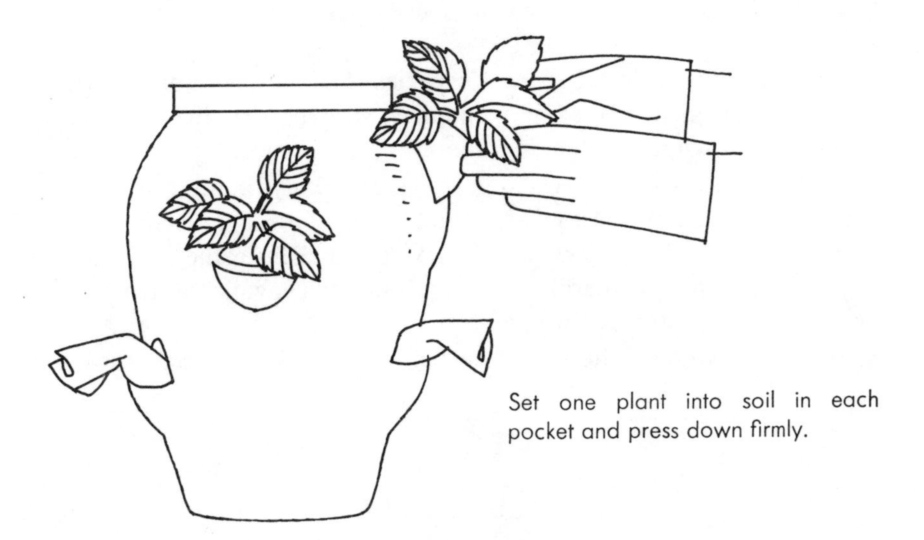

Set one plant into soil in each pocket and press down firmly.

planted all over it or can be interspersed with flowering plants such as impatiens for a very attractive display.

Still More Alternatives

Strawberries can also be grown on the patio in hanging baskets suspended from tree limbs, in tile flues filled with soil, and in decorated gallon cans attached to a fence. Old tires, scalloped along the edges with a sharp knife and colorfully painted, can serve as strawberry planters as well as flower patio planters—they might even be piled one atop the other and made into a strawberry barrel. For a really unusual effect, try inserting a few plants in soil-filled crevices of old logs. Or plant strawberries on the patio itself between bricks laid in sand. Another patio pleaser consists of setting individual plants in 5-gallon plastic containers that are filled with sterilized soil and set in the grass surrounding the patio.

Remember, too, that strawberries can be planted as a ground cover on front lawns if the soil is turned over for a year to kill grubs, and that the plants make a fine bankbinder for hills where it's hard to grow grass.

Domesticating Wild Gourmet Strawberries

"So plentiful and so sweet they make this a land of especial delight," an early settler wrote of the wild strawberries that "brightened whole New England hillsides and scented whole valleys." Often called "meadow berries," this species still persists untamed in rural areas throughout the country, the little berries a forager's delight and their leaves so fragrant that their hiding places among other plants in the shade of pine or beech woods can often be smelled out. Wild strawberries are, in fact, responsible for the strawberry clan's Latin name of *Fragaria* (fragrant) that Linnaeus gave it over two centuries ago, although the fragrance has been bred out of today's garden strawberries.

Wild strawberries consist of some 35 species, but only four or five were prominently involved in the breeding of modern strawberries and only three of these are much grown today. The 5 important species are:

FRAGARIA CHILOENSIS. Discovered in Chile, but really found from Alaska to Patagonia, this is a low, bushy plant with a pale-bluish underside to its green and glossy leaves. Its fruit is large, dark red, and firm, the hulls large. It was widely used in strawberry breeding, but isn't available from any nursery in the United States or Europe.

FRAGARIA VIRGINIANA. The wild strawberry of North America is a low-growing plant 4 to 8 inches high. Much used in strawberry breeding, the wild strawberry has small, very sweet fruit, but its little hull makes it difficult to prepare. This berry has to be gathered in the fields, for, to my knowledge, no nurseries offer it in the U.S. Unlike most small strawberries,

Strawberries in a hanging planter.

it propagates itself by runners which can be collected from the wild and transplanted in the home garden. It is the sweetest-smelling strawberry of all and, when domesticated, the little berries can grow almost a quarter of the size of regular strawberries. Wild strawberries are acid-loving plants and do well cultivated under blueberry bushes. Soon their runners make a thick ground cover that turns a soft red in autumn. Thin out the plants to assure a good crop of the soft, sweet, rather shapeless fruit and cultivate them the same as any strawberry.

FRAGARIA VESCA. This is the famous French *fraise des bois* native to Europe and much cultivated there. Also widely used in strawberry breeding, especially in developing everbearers, it is a hairy plant 9 to 12 inches high with thin, light green leaves. Its fruit is small, rarely an inch long, and hemispheric or slightly elongated. *Fraise des bois* is offered by several nurseries. Excellent plants can be ordered from The Guilde of Strawberry Bank, Inc., 93 State Street, Portsmouth, New Hampshire 03801, or from White Flower Farm, Litchfield, Connecticut 06759. The last nursery offers plants from French stock called Charles V, after the French ruler who in 1365 had his gardeners transplant wild strawberries from the woods and cultivate them in the gardens of the Louvre. Several nurseries also offer seed for propagating these everbearing plants, which bear their little berries from the end of June until frost. The runnerless plants are cultivated just like the Alpine strawberries, following.

FRAGARIA MOSCHATA. Sometimes called the Alpine Hautbois strawberry, the plant is very similar to *Fragaria vesca,* but taller and heavier. Native to Europe, it isn't much cultivated in the U.S., no nurseries offering it.

FRAGARIA VESCA SEMPERVIRENS. The Alpine strawberry, a variety of *Fragaria vesca* that originated in the mountains of Italy, is widely cultivated throughout Europe and America. Alpines, which gourmets consider the tastiest of strawberries, are available in five different varieties:

- *Baron von Solemacher.* The first widely available Alpine strawberry, this variety bears large red berries up to 1½ inches in length on plants about 8 inches tall. Dean's Improved Strain of Baron von Solemacher is a more rugged plant with the same-sized berries.
- *Harzland.* A good cropper that bears red berries about 1 inch long and ¾ inch in diameter.
- *Alexandria.* The newest variety of Alpine strawberry, an excellent cropper with the largest red berries of all on a plant that grows about ten inches tall and is very hardy.
- *Yellow Alpine.* Decidedly different for its pale yellow fruit, which has the same delicious flavor as other Alpines—even better, to some.
- *Cresta.* This plant is the only Alpine offered that reproduces itself by

Alpine strawberry.

runners. It has average-sized red fruit and handsome leaves variegated with white that make an excellent ground cover.

Starting Alpine Strawberries

Alpine strawberries of all types are usually grown from seed, which is available from many nurseries, including Parks, Burgess, Burpee, Shumway, Brecks of Boston, and Thompson & Morgan Inc. (see Appendix III for addresses). The seeds are generally sown indoors at least six weeks before the last expected frost. Since Alpine strawberry seed is very small and germination may be poor, take care in sowing it. Try sowing thinly in vermiculite, or in a thin layer of vermiculite spread over a pot filled with commercial potting soil. Or just press the seed slightly into fine soil and water carefully through a newspaper. Set in a sunny window, or under lights, the seed will germinate in from 10 to 14 days. Transplant the seedlings into peat pots when they make their first true leaves and grow them either in a window with at least a half day's sun, or under lights, feeding them biweekly with balanced water-soluble plant food at half strength. Plant the Alpines outside in early spring under the same conditions as you would regular strawberries. The plants, however, are so small that they are good for a rock garden and make neat flower-garden borders. Their small size also makes these highly ornamental plants well suited for container planting in hanging baskets, strawberry barrels, and strawberry jars, as described earlier.

Care and Propagation of Alpines

Alpines are hardy perennials, evergreen plants (except in the North) that bear their little berries all season long, and they usually fruit the same year that you start them from seed. They are very drought tolerant, especially if mulched, and can stand slight shade. The only real difference between Alpines and regular strawberries is that most Alpines do not propagate themselves by runners. The plants seed themselves in the garden and can also be increased by division every 2 to 3 years. Divide them in early spring. You'll notice that after a year or so plants will have two or three crowns. These are separated to double or triple the number of plants. As noted, Alpines are cultivated almost exactly the same way as regular strawberries, but the plant's leaves (*not* the crowns) are usually sheared every spring to increase productivity, while for bigger berries all but one or two trusses of flowers are picked out.

A dozen plants of Alpine strawberries will yield a decent harvest, but plant as many as you have room for—you'll never get enough of them. The gourmet berries slip off the calyx and don't have to be hulled like American wild strawberries. They keep well. Since the berry flavor is very concentrated, a small basket of Alpines seems a lot. Just a little sugar added to them a few hours before eating brings out all their delicious juices and they are better for jam making than garden strawberries because of their higher pectin content.

3

The Bountiful Blueberry

Blueberries have been a favorite food for centuries and are among the most widely distributed of all fruits. Early colonists here gathered the "blues," "whortleberries," and "bilberries," and made good use of them like the Indians before them. But cultivated blueberries more than any other fruit are children of the twentieth century. It was in the early 1900s that Elizabeth C. White of Whitesbog, New Jersey, one of several pioneer women fruit growers, offered prizes locally for highbush blueberries bearing the largest fruits. Hearing of her work, USDA plant breeder Dr. Frederick V. Coville began to work in cooperation with her, starting in 1909, and crossed many plants she or her contestants had selected from the wild in the Pine Barrens of New Jersey, an area with an acid, sandy, but fertile soil. By the time Coville died in 1937 there were thirty large-fruited, named highbush varieties where there had been none, and today there are myriad varieties that have been selected from hundreds of thousands of fruited hybrid seedlings. From its status as a lowly fruit often confused with the huckleberry (even though, unlike the bony-seeded huckleberry, its fifty to seventy-five seeds are small and barely noticeable when eaten), White and Coville had elevated the blueberry to a position where it became the basis for an entirely new agricultural industry. *Vaccinium corymbosum,* the highbush or swamp blueberry, had gone through a revolution rather than an evolution and became a mass-produced fruit in less than twenty-five years.

Today, only sixty-eight years after Coville released his first highbush variety, Pioneer, in 1912, thousands of acres are planted in blueberries in the United States. Five other species thrive here besides the highbush, making it possible to grow the fruit almost anywhere in the country. Unfortunately, however, many people have never tasted blueberries picked when ripe. Nearly 70 per cent of our $25 million crop of commercially grown blueberries are machine-picked as soon as they turn blue, even though they often need a week or more to become truly sweet, and these are what most people taste when they eat blueberries. Gardeners, discouraged by the myth that blueberries are extremely difficult to grow, miss out on a prime taste treat that is very nutritious as well—a pint of fresh blueberries contains 259 calories, 3 grams of protein, 64 grams of carbohydrates, 420 International Units of vitamin A, and 58 mg. of vitamin C. The berries themselves can be almost as big as grapes.

Actually, highbush blueberries are very easy to grow once their initial soil requirements have been fulfilled. They often need almost no care, and their disease resistance is incredible for a small fruit. Furthermore, with their pinkish bell-shaped flowers, red-toned stems and branches, and their glossy green leaves that turn red in autumn, they are a beautiful ornamental bush. Growing to a height of about eight feet, they can be used as lawn specimens, planted in shrub borders, or made into a fruit-bearing hedge far more attractive than privet. If you haven't tried growing them, treat yourself to a few bushes—six bushes, yielding up to 4 to 8 quarts each in their prime, should be more than enough for a family of four, and the bushes will live and bear up to forty years. Generally speaking, highbush blueberries do best in the eastern United States. Gardeners in really hot southern areas are better off planting rabbiteye varieties (see Index), but successful plantings of highbush blueberries exist from Maine to Georgia and from New Jersey with its warm, humid summers to the cool, dry summers of Washington's Puget Sound region. They have survived winter temperatures of 20° F. As a bonus blueberries attract songbirds, and many bird lovers grow them for that purpose alone—although birds, as we'll see, eat the berries so voraciously that they are probably the gardener's greatest enemy.

The Best Highbush Blueberry Varieties

One expert has noted that today's cultivated highbush varieties are "as different from their wild parents as modern hybrid corn is from its parents the Indians were growing when the white man discovered America." Highbush blueberries are bigger, sweeter, and more ubiquitous than any blueberry species—they are the kind you invariably buy when purchasing fresh blueberries in the supermarket, in contrast to the smaller lowbush kind, which are usually canned. Dr. and Mrs. Blueberry—Elizabeth White and Dr. Coville—selected many varieties from the wild, including Rubel, Dunfee, Harding, Grover, and Adams, and these were widely grown commercially, but now only Rubel of the wild selections survives. Most, if not all, of today's commonly grown varieties are the result of breeding work. All of them couldn't possibly be mentioned; for example, nurseryman J. Herbert Alexander, who introduced the excellent Herbert variety, is said to have developed over 270 other varieties in his fifty years' experience. I have included here the 28 varieties that are probably the most commonly offered by nurseries. To this list you might want to add Elizabeth, a midseason variety named after "Mrs. Blueberry"; Darrow, another midseason variety; Morrow, an early type; and Olympia and Pacific, two midseason types grown in the Pacific Northwest. If you can find it, Redskin, introduced in 1932 by the USDA, is an unusual red-skinned blueberry. Another interesting novelty, untested here, is the "Formosan Blueberry Tree" offered by Lakeland Nurseries, a *Vaccinium* species that grows in tree rather than bush form.

When buying blueberries be certain to buy at least two varieties (three is better) for cross-pollination. Most experiments have confirmed Dr. Coville's early report that "When blueberry flowers are pollinated with pollen from their own bush, the berries are fewer, smaller and later in maturing than when pollen comes from a bush of another variety." Most commercial growers, in fact, set out two rows of one variety and two rows of another. Blueberry plants generally begin bearing a little at 3 to 4 years old and yield full crops by the time they are 6 to 8. While most experts advise planting 2 to 3-year-old bushes that are 1 to 2 feet tall, and most nurseries offer this size, I have planted jumbo, specimen blueberry plants up to 8 years old and encountered no trouble at all; there is little waiting for fruit this way, but the bushes are, of course, much more expensive. No matter what size bush you buy, always try to buy blueberry plants with a burlapped ball of soil around their roots, which are very fine, fibrous, and hairlike in structure and should never be allowed to dry out. *The varieties asterisked here are the so-called "Top Ten" most often recommended by experts:*

Early Season Varieties
(Bearing from June to July)

ANGOLA. A bush grown mostly in North Carolina that is the earliest of all. Vigorous, spreading, and productive. The dark blue berries are medium-sized, round, somewhat soft, with good aroma and flavor, of excellent dessert quality.

*BLUETTA. A hardy, vigorous grower with a low compact habit, this new variety is offered by Kelly Brothers. Firm, light blue fruit that ripens very early and has fine flavor.

CABOT. An old standard that has been bettered by other early varieties and is no longer recommended. It ranks "medium" in about everything from productiveness and hardiness to fruit size and quality.

CROATAN. A vigorous, spreading, productive bush with medium to large, round, dark blue fruit that is firm, slightly aromatic, and of good dessert quality.

*EARLIBLUE. Vigorous and upright, this productive bush yields large, good-quality fruit resistant to cracking. It is an excellent early variety from Maryland northward that is larger, firmer, bluer, better-flavored, more rigorous, and more productive than Weymouth, which it is fast replacing as an extra-early variety.

JUNE. Bush of medium vigor, upright, and moderately productive, the leaves susceptible to a leaf spot that reduces plant vigor. Berries medium-sized, round, dark blue, firm, and of fair quality. Has been largely replaced by Earliblue.

MURPHY. A vigorous, productive, spreading bush with medium-sized round to oblate dark blue berries that are firm and only of fair quality. Canker resistance makes it a good berry for North Carolina.

RANCOCAS. Erect and productive bush of medium vigor. Berries small unless bush is pruned severely, oblate, firm, light blue, and of fair dessert quality. Leaves are subject to "June spot" dropping in the summer in Michigan, and the berry cracks badly after rain. In most respects Rancocas is inferior to the newer varieties.

WEYMOUTH. This erect bush is below average in vigor, with medium-sized, dark-blue, round to oblate berries of poor dessert quality. Besides the berries dropping, the plant is subject to canker, stunt, and mite injury. Earliblue is far better in the North and Wolcott in the South.

WOLCOTT. A very vigorous, upright, productive bush with medium-sized, round, dark blue berries of medium dessert quality. The variety is canker-resistant and well suited to the South.

Midseason Varieties
(July–August)

* BERKELEY. A vigorous, spreading, productive bush with firm, light-blue, oblate berries of good, mild, subacid flavor that are the largest of all blueberries. Is resistant to cracking and stores well.

* BLUECAP. Bush upright, of average vigor, hardier and more drought-resistant than most species, and productive to the point of overbearing. Berries large, oblate, of the lightest blue excluding Berkeley, firm, resistant to cracking, and of good quality.

* BLUERAY. Very hardy, upright, productive bush with very large, firm, light-blue berries that are highly flavored and resistant to cracking.

COLLINS. A vigorous, erect, moderately productive bush with large, firm, light-blue berries of good dessert quality.

CONCORD. The bush is vigorous, upright, hardy, and productive, bearing medium-sized, oblate, firm, light-blue fruit of good quality.

IVANHOE. Very large, firm, round to oblate light blue fruit that is one of the best in quality. A vigorous, erect, productive bush, but not reliably hardy north of Maryland and Delaware.

PIONEER. No longer much grown because it has been replaced by newer, better varieties, but a nice curiosity for the garden, because it was the first variety developed by blueberry pioneer Dr. Coville. The medium-sized, oblate, dark-blue berries do have a good flavor. The bush is of medium vigor, low, spreading, hardy, and productive.

SCAMMELL. Bush erect, vigorous, productive, and resistant to canker. Berries medium-sized and of medium quality.

* STANLEY. A vigorous, upright, hardy, and moderately productive bush with few main branches. Berries are oblate, light blue, firm, and sweet. They are large at the start of the season, but become too small at later pickings. Best of all in taste according to oldtimers.

Late Varieties
(August–September)

ATLANTIC. Bush very vigorous, spreading, and very productive. Berries exceptionally large, oblate, five-sided, light blue, firm, and somewhat acid if picked too early but good if harvested ripe. Fruit is resistant to cracking, but bush is subject to bacterial dieback in Oregon and Washington.

BURLINGTON. A vigorous, hardy, upright bush that is moderately productive and yields small to medium, firm, round to oblate berries of medium dessert quality. Resistant to cracking and has good cold-storage qualities.

* COVILLE. A very vigorous, spreading, productive bush bearing extra-large, round to oblate, firm, light-blue berries of excellent dessert quality when fully ripe. Berries hang on plant in good condition until early September.

* DIXI. Large, firm, oblate, five-sided, medium-blue berries of high dessert quality on a vigorous, spreading, productive bush. Bush is not very hardy in severe winters, and fruit is subject to cracking.

* HERBERT. Berries, among the largest and best in quality, are firm, oblate, and medium blue. Bush vigorous, upright, and productive.

* JERSEY. An old standby that was long everybody's favorite late variety and may still be the most widely grown of any variety. Bush vigorous, erect, very productive, and resistant to stem canker. Berries large, roundish to oblate, uniform in size, light blue, firm, acid when picked early but sweet when fully ripe. Bush is one of the hardiest and fruited well at the Geneva, New York, experiment station after winter temperatures that destroyed the crop and injured the wood on all other varieties grown there but Rubel. Jersey also produced a fair crop following a temperature of 20° F. on May 11, 1949, which destroyed the crop on other varieties except Rubel.

LATE BLUE. A new variety offered by Kelly Brothers that ripens late in August. A vigorous, productive bush bearing round, light-blue fruit of good quality.

PEMBERTON. An unusually vigorous plant that often comes into bearing a year or two earlier than most. Bush upright, hardy, and very productive. Berries very large, oblate, medium blue, firm, and of good quality. Fruit difficult to pick and subject to cracking in wet weather.

RUBEL. The only one of Mrs. White's selections from the wild still offered by nurseries. A small but excellent berry which, of course, has that wild blueberry taste and is of good dessert quality. Bush erect, vigorous, productive, and very hardy (see Jersey). Rubel is a parent of many, many blueberry varieties, including Jersey, Stanley, and Pemberton.

highbush blueberry varieties ranked for principal characteristics

SEASON (*early to late*)	SIZE (*lg. to sml.*)	COLOR (*light to dark*)	QUALITY (*good to poor*)	SHAPE (*erect to spreading*)
Angola	Berkeley	Berkeley	Dixi	Rubel
Weymouth	Coville	Bluecrop	Ivanhoe	Rancocas
Earliblue	Herbert	Blueray	Herbert	June
Bluetta	Blueray	Earliblue	Bluetta	Collins
June	Dixi	Bluetta	Blueray	Pemberton
Murphy	Ivanhoe	Stanley	Stanley	Scammell
Cabot	Collins	Collins	Croatan	Earliblue
Croatan	Atlantic	Ivanhoe	Earliblue	Bluetta
Rancocas	Late Blue	Jersey	Collins	Stanley
Wolcott	Pemberton	Atlantic	Pioneer	Late Blue
Ivanhoe	Croatan	Concord	Bluecrop	Coville
Blueray	Earliblue	Late Blue	Coville	Wolcott
Stanley	Bluetta	Burlington	Late Blue	Ivanhoe
Collins	Bluecrop	Rubel	Atlantic	Jersey
Concord	Angola	Cabot	Wolcott	Bluecrop
Bluecrop	Weymouth	Coville	Concord	Berkeley
Pioneer	Wolcott	Scammell	Murphy	Herbert
Scammell	Jersey	Rancocas	Berkeley	Concord
Berkeley	Murphy	Herbert	Pemberton	Blueray
Atlantic	Concord	Wolcott	Angola	Burlington
Pemberton	Stanley	Dixi	Burlington	Dixi
Rubel	Burlington	Pemberton	Jersey	Angola
Herbert	Pioneer	Pioneer	Rancocas	Atlantic
Jersey	June	June	Scammell	Croatan
Dixi	Rancocas	Murphy	June	Weymouth
Burlington	Rubel	Weymouth	Rubel	Murphy
Coville	Scammell	Croatan	Cabot	Pioneer
Late Blue	Cabot	Angola	Weymouth	Cabot

Seven Ways To Make "Blueberry Soil"

Before your order of blueberry plants arrives be sure the soil is ready to receive them. The main reason blueberries weren't cultivated until this century is that gardeners failed to realize that the bushes had to be planted in acid soil. If you have soil where azaleas, rhododendrons, camellias, laurel, holly, or any acid-loving plants thrive, blueberries will do excellently there. If not, it is best to change the soil.

Blueberry soil should have a pH of from 4.2 to 5.2—anything below that range is too acid and anything above it is too alkaline. First select your planting site (see the Index) and then either test your soil with an inexpensive soil-testing kit or take samples from around the planting site, mix them together, and have them tested by your state agricultural extension service. Very few soils will test out too acid and those that prove too alkaline can be changed by *one* of the following methods:

- Dig a trench about 2 feet deep, 3 feet wide, and as long as you need to accommodate your plants set 4 to 5 feet apart. Fill the trench with a mixture of 2 parts builder's sand, one part good garden loam, and 2 parts acid peat (such as sphagnum moss), which can be purchased from garden centers or dug from a peat bog if you are fortunate enough to have one nearby. This mixture should test out below pH5.0, but if it is higher, add more acid peat. Plant the blueberries in this trench (see Index) and mulch them (see Index).
- As an alternate to the above, dig the same trench, and fill it with good garden loam, builder's sand, and any of a number of acid materials such as shredded or rotted oak leaves, pine needles, leaf mold, and woodland soil, sawdust, or wood chips. (*Do not* add manure, which makes the soil alkaline.) Then mulch the plants. Least expensive of any mix would be equal parts of good garden soil, rotted oak leaves, and builder's sand (or even beach sand washed free of salt). You need only be sure that any mixture tests out to between 4.2 and 5.2.
- Forgo the trench system and just dig a hole 2 feet deep and 3 feet across for each blueberry bush you plant. Fill the planting hole with any of the above tested acid-soil mixtures and mulch as above.
- Cut a 50-gallon metal drum in half, punch a half-dozen drainage holes in the bottom, and sink it into the soil, leaving its rim protruding several inches above the soil line. Fill the drum with any of the acid-soil mixes above and plant one bush in it. Again, mulch as above. This is an especially

good method for areas with very alkaline soils (over pH7.0).

• Correct the soil by chemical means with sulfur or aluminum sulfate. Sulfur is probably the cheapest of these two acidifying agents. Sandy soils need ¾ pound of sulfur added per 100 square feet for each full point that the soil registers above pH4.5, medium loams need from 1½ to 2¼ pounds, and very heavy clays do not respond well to such treatment. Work the sulfur into the soil to a depth of 6 inches. Do this early in spring, well before planting, in order to allow enough time for soil bacteria to oxidize the sulfur, as large amounts of the substance injures plants when applied directly to the roots. Organic growers do not use sulfur or aluminum sulfate, as they believe that they (especially aluminum sulfate) make the flavor of the fruit acid and are detrimental to bacterial life within the soil.

To correct soils that are *too acid* (below 4.0) for blueberries, which isn't often a problem, simply add enough agricultural-grade limestone to bring the soil as close as you can get it to the optimum level for blueberries. Do your liming before setting out the plants in the spring—preferably the previous fall. Aside from correcting overacid soil, do not use lime on blueberries at any other time—*ever.*

Planting Blueberries

Where to Plant

In addition to an acid soil, blueberries require an open, porous soil that provides many minute passageways for their fine roots. Sandy soils, peats, and loams are all fine, but compact, heavy clays that the roots cannot push their way through must have sand added to them. Blueberries can't withstand drought, either, and need a moist soil (ideally one where the water table is 14 to 30 inches below the surface), though not one that is waterlogged—good drainage is essential. A soil that yields plentiful vegetables would satisfy these requirements admirably, but pastureland that is too wet for other crops is good for blueberries, too. It is important that lots of organic matter be incorporated into the ground to provide aeration. Compost can be worked into the soil before planting, or a green cover crop can be planted the previous autumn and turned under before setting out the blueberry plants.

Blueberries should be planted in full sun in an open location that is free of grass and weeds at least 3 to 4 feet from the plants. At least two-thirds sun is necessary for a good set of bloom and fruit, although the plants will yield some fruit in half sun and do reasonably well with only 3 to 4 hours of sun a day on them. Slightly sloping land or a level area not surrounded by higher land or trees makes for an ideal location with good air circulation. Avoid both high, dry places and frost pockets—low areas surrounded by higher land from which cold air descends on frosty nights, making minimum temperatures several degrees lower. Areas surrounded by trees and brush should also be avoided, as their reduced air circulation makes them more subject to severe frosts than open areas. Poor air circulation also prevents berries from drying after rains, which encourages the growth of mummy berry fungus, a disease that causes them to dry up and drop off before harvest time.

Since low-lying areas are usually selected for blueberries, make certain that such sites are provided with proper drainage if they do not have it. Blueberries are often drowned when planted in depressions or pockets where water doesn't drain out. The bushes need plenty of water, but not where it lies on their roots during the growing season (water on their roots when they are dormant doesn't seem to bother them).

When and How to Plant

Planting is best accomplished when blueberries are dormant in early spring or late fall, though balled and burlapped specimens can be safely planted almost any time. Spring planting should be done as early as the soil can be worked and fall plantings can be made from the time the leaves begin to turn color until the ground freezes. In fall planting it is best to mound soil up around the bushes a few inches or to use a mulch. This prevents the roots from heaving out of the ground during the winter due to alternate freezing and thawing. As noted, order balled and burlapped plants whenever possible. Take every precaution not to let blueberry roots dry out if the plants aren't balled and burlapped. Barerooted plants should be heeled into a shallow trench in a shady place until they can be planted and kept covered from the sun and wind when they finally are set out.

Have a hole 3 feet wide and 1½ feet deep ready for each blueberry bush before planting—large enough to allow the roots to spread out rather than hang down. Plant the bushes just a trifle deeper than the soil line on the main stem, fill in the hole, and water well. Never fertilize when setting out a blueberry bush—wait until it has leafed out (see Watering and Fertilizing, in the Index), but do put down a mulch right away (see Mulching Blueberries, in the Index).

The bushes can be planted anywhere from 4 to 6 feet apart from one another each way, depending on how much room you have—the more the better, so the sun can reach all parts of each bush. There should be at least 8 feet, and preferably 10 feet, between rows for tractor cultivation if a large planting is planned.

Space Savers

If space is a problem, blueberries can be grown as lawn specimens, or made into an attractive front yard "living fence" to take the place of privet.

A flowering, fruiting blueberry fence is made quite simply by planting blueberries 3 feet apart instead of the usual 6 feet. Depending on the variety used, a fruiting fence can range from 3 to 6 feet high. Recommended varieties (and use several of them for good fruiting) are Earliblue, Blueray, Berkeley, Herbert, Coville, and Jersey. Set and care for the plants as you would any blueberry.

Blueberries are also easy to grow on city penthouse terraces and suburban backyard patios. Use containers no less than 24 inches in diameter and height, punching several holes and adding gravel in the bottom for drainage and filling them with any of the acid-soil mixtures noted in Soil for Blueberries (see Index). Mulch the plants and care for them just as you would for blueberries in the garden, and be especially careful not to let them dry out, even if it means watering every day in the summer. Additionally, it helps to root-prune potted blueberries every 2 to 3 years by digging deeply into the tub and severing and removing several bunches of the fibrous roots.

Care Of Blueberries

Mulching and Cultivating

Home-grown blueberries don't have to be cultivated if they are mulched, for a proper mulch will keep down weeds sufficiently and conserve soil moisture. Mulches also help fertilize blueberry plants and keep the soil acid. In fact, the shallow-rooted blueberry really can't be grown well without mulches—especially in the South, where mulches hold down soil temperatures.

Mulches for blueberries include just about any nutritious and acid material. Sawdust, wood chips, oak leaves, hay, straw, acid peat, pine needles, seaweed, or even spent hops from a brewery are fine. Mulch each plant with 4 to 8 inches of whatever material you choose, but don't smother the stems with the material, leaving a clear space of 4 to 5 inches between the mulch and stems. Mulch between plants and between plant rows, too, filling in at least 2 feet on each side of every plant. Be sure to apply the mulch only after the soil is full of moisture, preferably after a heavy rain, and keep replenishing the mulch, as it decomposes every year. If you want to, you can slip rubber tires over your plants and fill them with mulch.

In order to ensure blueberries a chill period when they are dormant, mulches can be pulled away from the plants during winter, a practice I've never found necessary. Sawdust is the mulch most often recommended, but I've found it to be the least satisfactory material, for it packs too much, often deprives plants of nitrogen, turning leaves yellow, and has to be supplemented with a nitrogenous fertilizer. In my own experience I've discovered that an 8-inch shredded-oak-leaf mulch works best, and since using it I've never had to weed or water my bushes. (See also Appendix II, Fifty Mulches for Berries.)

Watering and Fertilizing

Mulching will usually take the place of watering where blueberries are concerned, but check the plants periodically, never allowing the soil to dry out.

Fertilizing your blueberries may not be necessary, either, if the soil is rich and peaty enough. However, most gardeners do fertilize. Dr. Coville, the blueberry's great developer, used a mixture of dried blood, sodium nitrate, phosphate rock, potash, and steamed bone meal to triple yields of blueberries grown in sandy soils. Gardeners have since devised many methods and any one of the following would suffice:

- My favorite practice is to fertilize blueberries in March, April, and May with three small applications of organic sources of nitrogen, phosphorus, and potash in equal amounts. (Natural sources of nitrogen include dried blood, cottonseed meal, fish meal, bone meal, and manure.) About three handfuls of the mixture are enough for a 3-year-old bush and it should be spread around the drip line of the bush, away from the main stem. Use more fertilizer every year as the bush grows, until by the time it is 8 to 10 years old it is receiving about 2 pounds. Making three small separate applications of fertilizer helps fight the temptation to overfertilize and it is better to err on the lean side in fertilizing blueberries than to overfertilize, since too much nitrogen produces beautiful foliage and no fruit. Fertilizing in March, April, and May will give your plants food at the time they most need it—when they are producing new feeding roots, developing fruit buds set the previous year, and setting fruit buds for next year.
- Some gardeners simply use a natural nitrogenous fertilizer like cottonseed meal, spreading ¼ pound around young plants and double that around old ones early in spring.

• Other gardeners fertilize with ammonium sulfate. As noted, organic gardeners don't like to use this or any chemical fertilizer, believing that it makes fruit flavor acid and is detrimental to the soil. A maximum of ¼ pound of ammonium sulfate would be used for a plant 5 to 6 feet tall, and it should be applied in early spring around the drip line of the bush.

• Another method is to fertilize with 10-10-10 chemical fertilizer, using 1½ ounces for every year the plant has been in the garden, but no more than 8 ounces.

Whichever method you choose to fertilize your plants, remember *not* to fertilize when setting the bushes out and never to fertilize after June 1. Late fertilizing of blueberries only stimulates late shoot growth that won't be able to harden up for winter.

Pruning

Blueberry bushes don't need much pruning the first few years except for cutting out twiggy growth at the bottom and picking off all blossoms the first 2 years to allow the plant to strengthen. After the plants are 4 to 5 years old, pruning should be done annually in the dormant season, either in late fall,

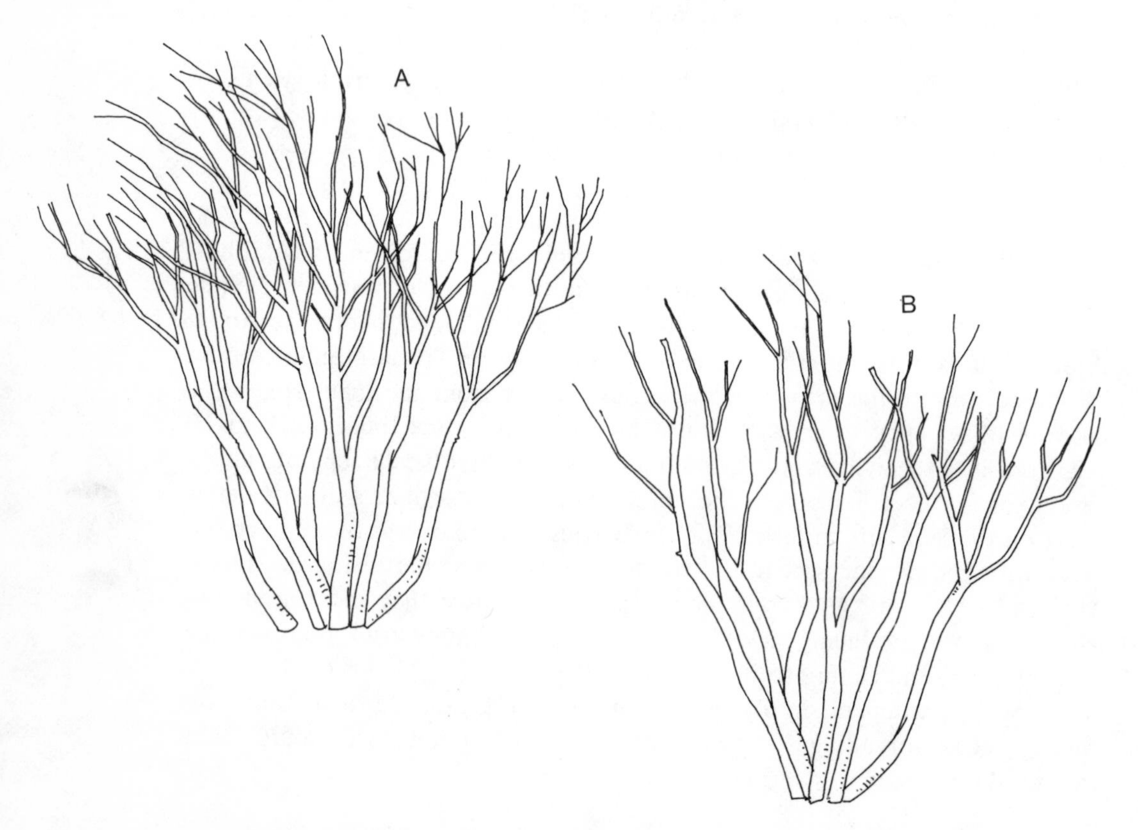

A four-year-old blueberry bush showing how it should look before (A) and after (B) pruning.

winter, or early spring, when the plump fruit buds can be distinguished from the thinner leaf buds. A good rule of thumb to remember is that light pruning will obtain larger harvests of smaller berries that ripen later, while heavy pruning obtains smaller harvests of bigger berries that ripen earlier. Much

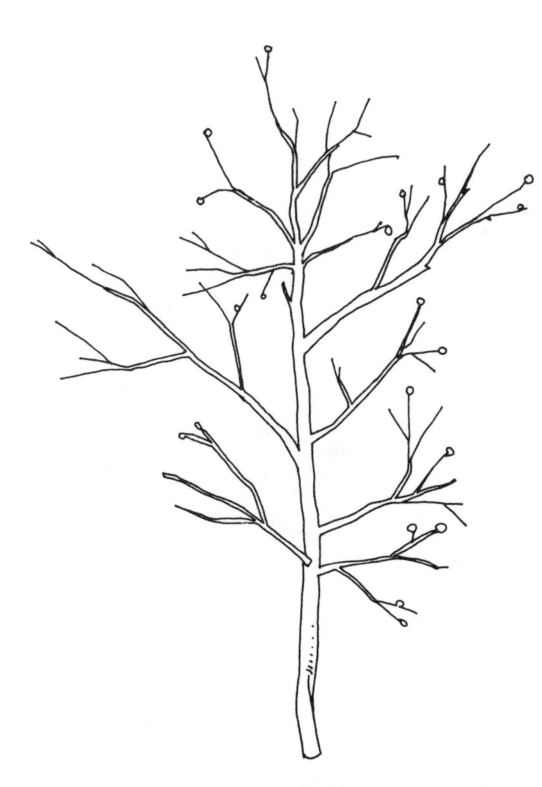

This blueberry cane had too many fruits last year, resulting in poor shoot growth. It should be headed back to the strong lateral and some of the twigs should be removed.

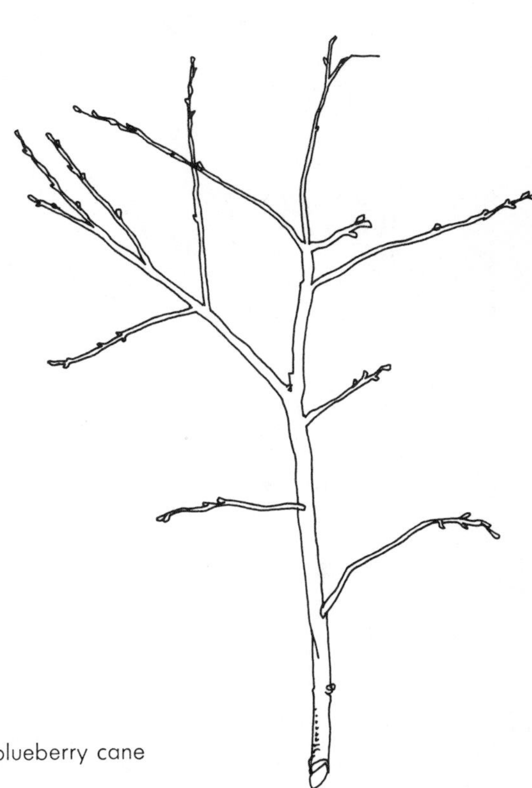

A vigorous, well-branching blueberry cane that needs no pruning.

depends on the vigor of your plants, which you'll have to observe, but the following is a general guide to annual blueberry pruning:

1. Limit the number of canes arising near the crown to one for each year of age of the bush (or one for each foot of its height), up to a maximum of 6 to 8 canes for bushes over 8 years old.
2. Prune out all canes over 6 years old.
3. Remove sucker shoots, all weak twiggy branches, and soft short shoots less than 3 inches long.
4. Cut back the tips of vigorous fruiting wood to about one fruit bud per 3 inches of shoot growth. This will leave 4 to 5 fruit buds on each twig. Each fruit bud yields a cluster of up to 14 berries.
5. Cut back all diseased or damaged wood to healthy wood.
6. Collect all prunings and dispose of them in a sanitary manner, since they spread disease.

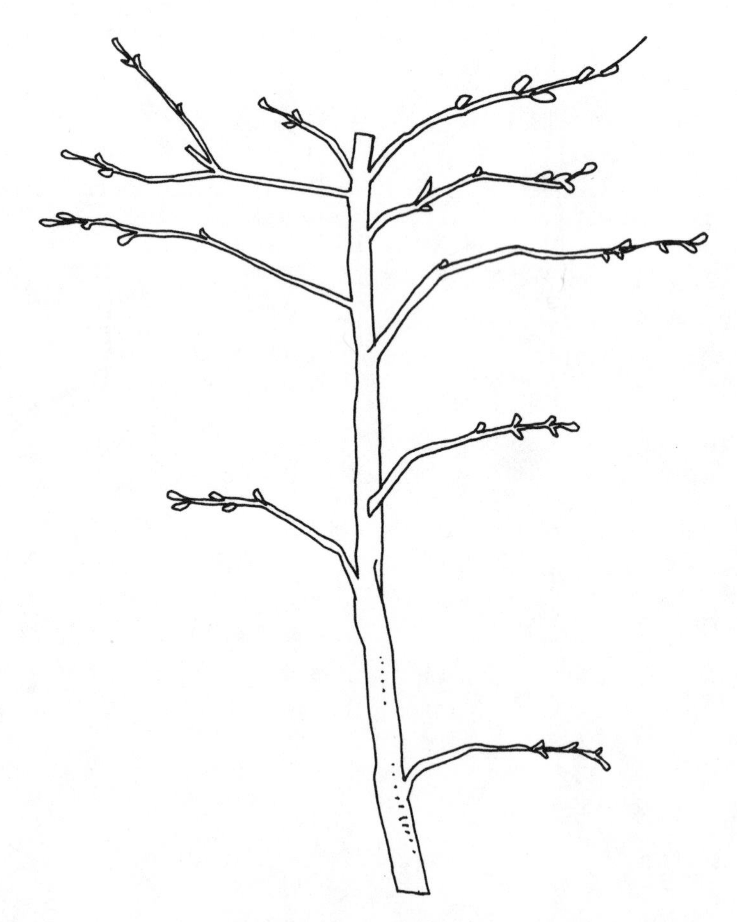

Heading back a vigorous blueberry cane forces out many strong lateral or side branches that bear large fruit.

Diseases and Pest Troubles

I have grown blueberries off and on for over fifteen years and never encountered insect or disease trouble with them in all that time. This is not unusual among home gardeners; only large-scale commercial growers seem to report major blueberry disease or pest troubles. You'll keep blueberry damage to a minimum by pruning all diseased wood, disposing of all trimmings after pruning so that they don't provide a breeding place for insects, picking all fruit from branches, and generally keeping the garden clean. The odds are that you'll never have to spray the bushes and there's a good chance you will never experience any of the following most common blueberry troubles:

BIRDS. Can easily eat all the fruit in a planting if not deterred. Use any of the protections noted in Appendix I, For the Birds, especially netting the bushes with cheesecloth or plastic.

BLUEBERRY MAGGOTS. The blueberry fruit fly lays eggs in ripening blueberries and the larvae or grubs hatched infect the berries, rotting them. It is important to clean dropped berries from the ground under blueberry bushes, for these might contain maggots that will keep the cycle going by wintering over in the soil. A fairly good control is to dust with the safe insecticide rotenone when berries begin to turn blue.

BLUEBERRY STEM BORER. Bores through canes, causing wilting. Best control is to prune off the infected cane well below the affected area. Some growers spray with Sevin in May.

BOTRYTIS BLIGHT. This wet-weather blight causes berries to shrivel and turn purplish and causes shoot tips to die. There is no effective control, but avoiding fertilizers high in nitrogen helps, as does regular pruning.

CHERRY FRUITWORM. A small red worm that feeds on fruit and can be controlled organically by the parasite fungus *Beauveria bassica.*

CRANBERRY FRUITWORM.

FALL WEBWORM. A caterpillar that spins a web on plants in early August. Hand-destroy webs and caterpillars.

JAPANESE BEETLES. Eat leaves of plants. Milky spore disease preparation is a good organic weapon to use against them. Commercial traps are also available.

MUMMY BERRY. A fungal disease that causes berries to shrivel, turn brown, and harden, and kills tender shoots. To control, cultivate the ground around the bushes to a depth of 1 to 2 inches in early spring and do not overfertilize.

PHOMOPSIS TWIG BLIGHT. This disease girdles and shrivels young shoots; the only control is to prune off all infected growth.

POWDERY MILDEW. A white mold on the upper part of the leaves that appears after harvesting and probably does the plant little or no harm. There is no control, but a few varieties like Earliblue and Ivanhoe are resistant to powdery mildew.

STEM CANKER. Once very prevalent in the South, this disease eventually kills plants. Stem canker isn't much of a problem anymore thanks to canker-resistant varieties like Angola, Croatan, Murphy, and Scammell.

STUNT DISEASE. A virus disease that is spread by leafhoppers, which can be controlled by spraying with Sevin. Symptoms are little growth of plant and very small leaves, unproductive bushes. Stunt, however, isn't a serious problem anymore, even in the South, for virus-free plants are widely available from nurseries. If plants become infected with this disease after planting (catching it from wild blueberries in the area), destroy the stunted plants, which will never be normal again.

YELLOWING OF LEAVES. Called chlorosis, this condition is associated with a too alkaline soil, resulting from a lack of iron in the soil, but it can also be caused by poor drainage. If the former is the cause, apply iron chelate to the soil. GU49, available at garden centers, is particularly good.

PROPAGATION

Blueberries are most often increased from *hardwood cuttings,* which is a relatively simple process. Most varieties root easily, and if you're patient enough to wait the 4 to 5 years it takes for a cutting to become a producing blueberry bush, this method is excellent. In late winter or early spring just

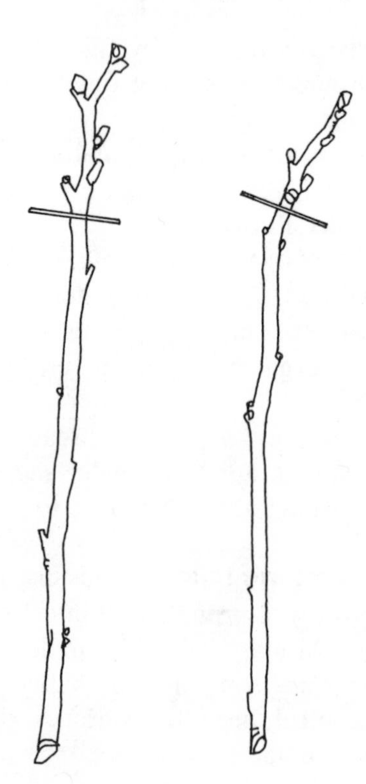

Hardwood blueberry cuttings with fruit buds at the tips. Cut them as indicated before planting.

take 3 to 6-inch cuttings from the dormant shoots of the previous year's growth. Try to take cuttings that have only thin leaf buds (they root better than cuttings with fatter fruit buds) and insert them in a mixture of half acid peat moss and half sand, leaving only the top bud sticking out above ground (at least two thirds of the cutting should be covered). Don't let the cuttings dry out (but don't overwater them either) and make sure that the greenhouse or outdoor propagating frame is properly ventilated so that the cuttings have good air circulation. The cuttings that root can be transplanted to a nursery bed the next spring and set out in the garden the following year. If they are propagated in a cold frame, cover them with straw mulch over the next winter to prevent alternate freezing and thawing of the soil.

SOFTWOOD CUTTINGS, not usually so successful, are grown in essentially the same way. They are taken when secondary growth first appears on the new shoots in spring and are made about 4 inches long. Two leaves are left on each cutting here, too, but the upper half of each leaf is also cut off to reduce transpiration. Softwood cuttings are generally used to propagate southern rabbiteye varieties (see Index), which are difficult to root by hardwood cuttings.

MOUNDING is another good blueberry-propagation method. Here acid soil or peat is piled a foot or so high around a bush whose canes have been wounded where they will be buried beneath the mound. The wounded canes will root and by the next spring can be cut off below their roots and planted in the garden.

An interesting and very easy way to propagate blueberries and obtain large plants is the *air-layering method,* which is rarely used by gardeners. While the standard equipment for air-layering (offered in kits at garden supply stores) includes a sharp knife, rooting stimulant, sphagnum moss, polyethylene plastic sheeting, and plastic electrical tape, I found that just the sharp knife was absolutely essential by using only materials that I had on hand. For sphagnum moss, I substituted acid peat moss in one case, black peat in another, and rich organic garden soil in the last. The special plastic sheeting was dispensed with in favor of plastic freezer bags for two layerings and burlap for the other. No rooting stimulant was needed and, instead of plastic electrical tape, I used black friction tape or string.

Since I was breaking "laws," I decided not to use year-old branches, either, but selected older limbs on three 8-year-old Jersey plants—older limbs usually take longer to root if they root at all. What's more, layering was done late in July, not in cool weather when rooting is most vigorous and air-layering is usually practiced.

The layers were made the proper 18 inches from the tip of each branch, all leaves within 6 inches being removed. Instead of wounding the wood by slitting or notching, I ringed or girdled the branches a half inch, scraping lightly to remove the slippery cambium, and squeezing the moist peat or soil around each wound to remove excess water. Then the bags and burlap were fixed securely in place with tape or string.

The whole operation took about 15 minutes, and I had little hope for its success, but knew there was nothing to lose except three branches that would have been pruned off anyway. I did watch the wrappings to see that no water seeped in, and made sure that the burlap didn't dry out, but otherwise forgot about the whole matter until late fall. Air-layers started in spring frequently do not root until the next spring, but I discovered in midautumn, to my surprise, that all three had succeeded. Roots had formed on each of the injured branches and the rooting medium in both plastic bags was filled with fibrous roots, while the growth under the burlap wrap was just as thick.

There are two schools of thought about transplanting air-layers. One claims that the new roots will be injured by freezing if left on the plant over the winter, advising quick transplanting before cold weather. The other holds that the air-layer should be left unsevered from the parent plant for a full year. To test both theories, I transplanted the burlap-wrapped layer and one from a plastic bag to the nursery bed that fall, and left the third on the shrub until the next spring. Both methods worked—all three new blueberries are thriving today.

When air-layering more conventionally—that is, not using my method—notch any young branch 18 inches from its top, sprinkle the wound with rooting hormone, removing all excess powder by shaking the branch, and wrap the wound with moist (but not wet) sphagnum moss. Next fix polyethylene plastic in place with electrical tape. When roots form, the layered growth is removed and given tender care in the nursery bed until ready for transplanting.

Harvesting Blueberries

For sweeter fruit allow berries to hang on plants for at least one week after they turn blue before picking. Ripe berries will come loose easily and drop into your hand; if any pressure is needed to pick them, they will not be ripe. A reddish ring around the "scar" or indentation where the fruit is attached to the stem also indicates that a berry is unripe.

Pick fruit early in the day before the sun gets hot, as berries have more flavor at that time. Similarly, do not pick blueberries after a rain, as the flavor is often diluted with water at such times. Pick at weekly or ten-day intervals.

It is best to pick blueberries by gently rolling them from the cluster with the thumb into the palm of the hands; in this way the berries that are not ripe enough and are more firmly attached will not be picked. Never pull or force berries from the plant. A handy container to use is an empty Quaker

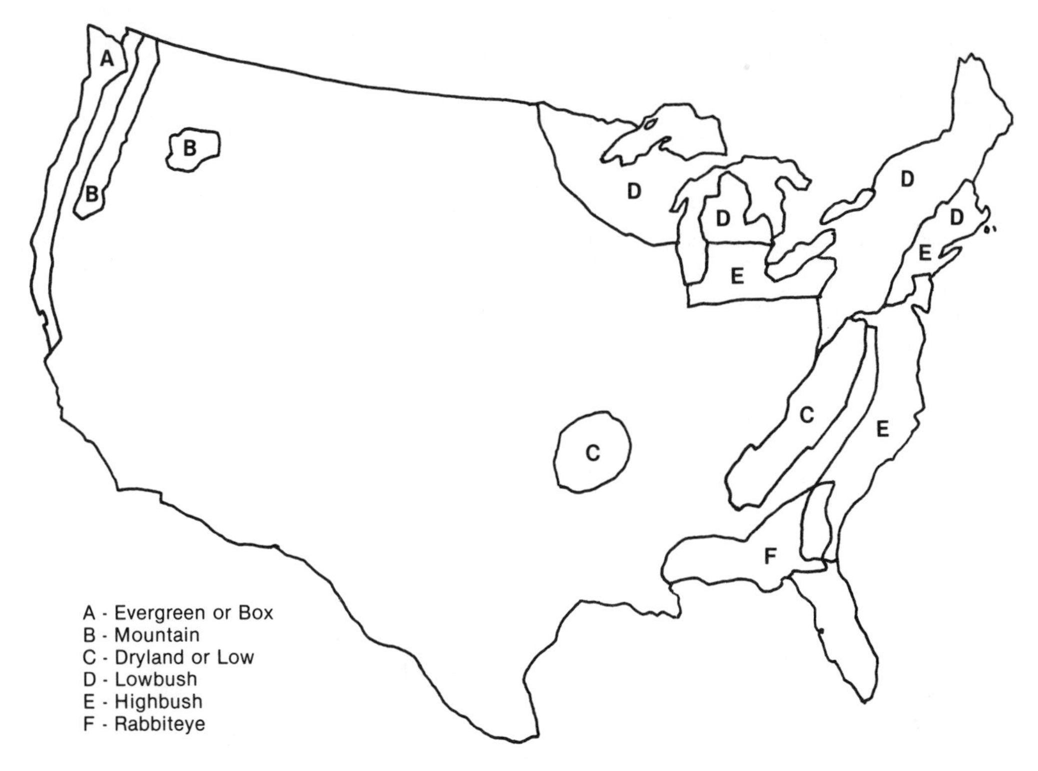

Map of the United States, showing areas in which wild blueberries are extensively harvested.

Oats box or milk carton strung through holes punched near the rim and hung around your neck.

Rabbiteyes: Blueberries For The South

In 1893, even before Elizabeth White and Dr. Coville began their pioneering work on the highbush blueberry in the New Jersey Pine Barrens, a grower named M. A. Sopp was establishing the first commercial planting of southern rabbiteye blueberries. The rabbiteye, so called because to some the berry resembles the eye of a rabbit, hadn't had much of a reputation because of the small, rather gritty black fruit most wild varieties produced. But Sopp and other growers selected excellent types like Clara, Walker, Black Giant, and Ethel from the wild, changing all this. Later, starting in 1940, the USDA began developing hybrids from such selections that are now almost as good as today's highbush varieties.

The rabbiteye (*Vaccinium ashei*) is a native of river valleys and pine barrens in southern Georgia and Alabama and northern Florida. In contrast to the highbush blueberry, which sometimes needs to be "chilled" 1,000

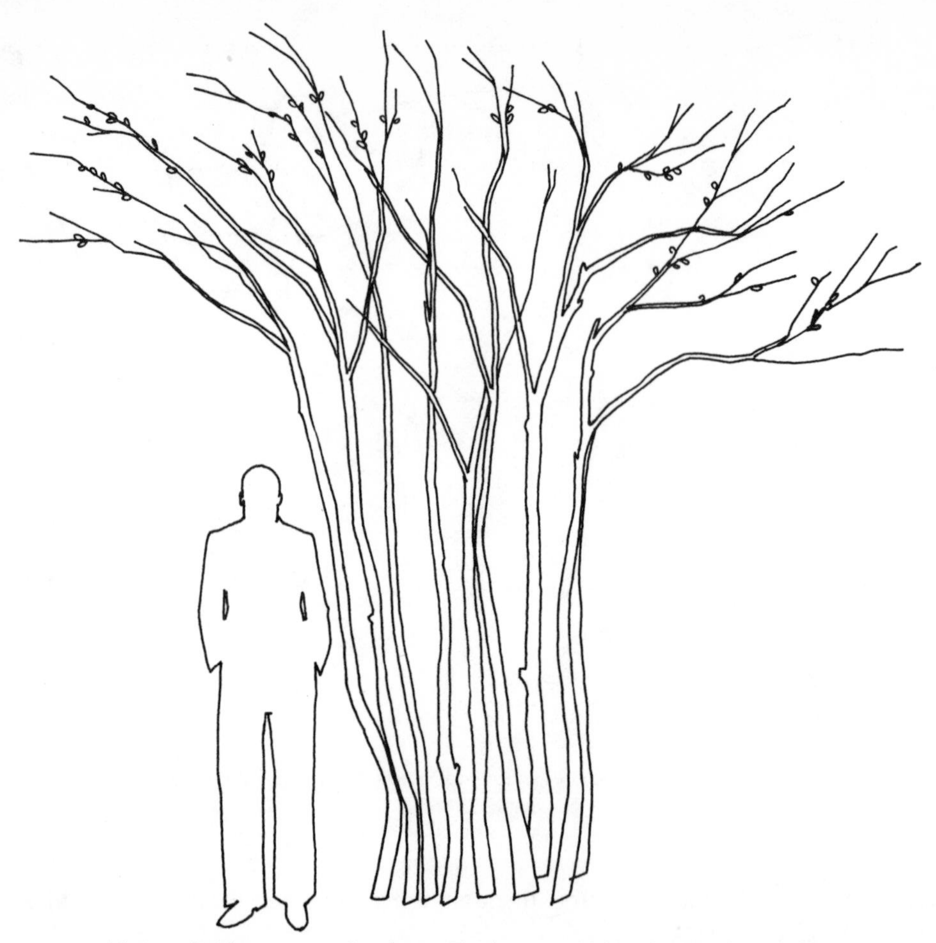

Mature Rabbiteyes or Southern Blueberries. Unlike highbush blueberry species, these grow to great height, often reaching 8 to 20 feet.

hours below 45° F. every year, it needs only 500 hours below 45° F. A plant that grows much taller than the highbush species, ranging from 8 to 20 feet tall, it is not so sensitive to soil acidity and is far more heat- and drought-resistant. Other big differences are that the rabbiteye needs much less pruning and produces many more berries—up to 30 quarts per plant on 11 to 12-year-old bushes.

Rabbiteyes should definitely be the blueberry choice of most gardeners in southern and Gulf Coast states. Grow the rabbiteye essentially the same as you would a highbush blueberry. However, plant it from 8 to 15 feet apart each way, according to bush size (see chart following), and no less than 8 × 8 feet in the home garden. Don't be afraid to plant in less acid soils and well-drained uplands—although the rabbiteyes ideally prefer conditions as similar as possible to those liked by the highbush. About the only pruning you need do is to remove dead or damaged wood and perhaps thin out the shoots on older plants every few years. This species is even more resistant to disease and insect pests than the highbush and isn't bothered at all by stem canker, that bane of highbush blueberries in the South. Rabbiteyes are generally propagated by suckers or softwood cuttings, not by hardwood cuttings like the highbush blueberry (see Index).

Rabbiteyes bear from June through July and their berries hang on the bush for a week or more without harm. They can be used for the same recipes as highbush blueberries, though their flavor isn't quite as good yet. I say "yet" because in the future some cross between rabbiteyes and highbush varieties is bound to produce a southern variety equal in taste to the northern ones. In the meantime try any of the following rabbiteye varieties, which are ranked below according to various characteristics. Other good varieties not rated here include Clara and Black Giant, selections from the wild; Delite, a new, late-ripening, light-blue variety; Southland, a late variety, with sweet, rich, blue berries; and Georgia 40. Be sure to plant more than one of any kind, as rabbiteyes are even more self-sterile than highbush blueberries. Plant the bushes an inch deeper than they grew in the nursery and shade the young plants with pine boughs or other material until they become established.

The Best Rabbiteyes

Rank	Season	Size of Berry	Dessert Quality	Color	Bush Size
	(early to midseason)	*(large to small)*	*(good to poor)*	*(light blue to black)*	*(large to small)*
1.	Woodard	Woodard	Woodard	Tifblue	Tifblue
2.	Coastal	Tifblue	Tifblue	Woodard	Homebell
3.	Callaway	Menditoo	Garden Blue	Garden Blue	Garden Blue
4.	Homebell	Coastal	Menditoo	Ethel	Coastal
5.	Garden Blue	Ethel	Homebell	Homebell	Woodard
6.	Ethel	Homebell	Callaway	Walker	Ethel
7.	Walker	Garden Blue	Walker	Menditoo	Menditoo
8.	Tifblue	Walker	Ethel	Coastal	Walker
9.	Menditoo	Callaway	Coastal	Callaway	Callaway

Lowbush And Other Unusual Blueberries

Fruit can be harvested from a number of blueberry species besides the highbush and rabbiteye types. None of these is extensively grown by home gardeners—indeed, no named varieties have been developed from any of these "unusuals." But all are worth a try, and if you don't want to plant them, you can always gather their delicious berries from the wild.

Lowbush

A plant well worth growing is the lowbush blueberry (*Vaccinium lamarckii,* formerly *V. angustifolium*) that is native to the northeastern United States and Canada. This is an upland berry also called the "low sweet" that is commercially grown in the sandy Maine blueberry barrens and is largely used for canning. The bluish-black fruits, which ripen in August, are smaller than those produced by highbush or rabbiteye species, but are extremely tasty, with a slight honey flavor. Lowbush blueberries are deciduous, boast delicate lily-of-the-valleylike flowers and flaming autumn foliage. Growing from 6 inches to 2 feet high and more difficult to domesticate than highbush species, they make an excellent ground cover and will thrive in just a few inches of soil.

The lowbushes are planted 2 × 2 feet apart in the home garden and need an acid soil made of ⅓ sand, ⅓ peat, and ⅓ forest duff—it is even more important to give them an acid soil than it is for highbush types, so definitely change your soil to "blueberry soil" (see Index) if it doesn't fit these qualifications. Care for the lowbushes just as you would for highbush blueberries and use the fruit the same ways. If you don't do much pruning, they'll grow to 18 to 24 inches; if you do prune regularly, they'll form a dense, compact stand or ground cover with larger fruiting clusters. The berries, low to the ground, are generally harvested with a blueberry rake.

Commercial growers usually "brown over" lowbush blueberry plantings every second or third year in the spring while the ground is still wet and no growth has yet begun. This burning of blueberry fields is facilitated by spreading straw or hay around the field and effectively keeps down underbrush, weeds, and insects, resulting in a more productive planting.

LOWBUSH BLUEBERRIES ARE PROPAGATED by taking a shovelful of what is called "blueberry sod" from a field where they are growing. The chunks of sod are transplanted and usually contain enough of the rhizomes or underground stems that the lowbush blueberry spreads by to produce new plants. (You can also dig rhizomes from a blueberry field in the fall, cut them into 4-inch lengths, and store them covered with moist moss in a cold place with a temperature no less than 40° F.) The buds on the rhizomes will send up vegetative stems and the rhizomes will be ready to plant when all danger of frost is past.

Rhizomes or softwood cuttings will produce true-to-form, lowbush blueberries that are exactly like their parents, but they don't spread as rapidly as lowbushes grown from seed. It's not easy to grow lowbush blueberries from seed and they do not come true to form, yet many growers prefer to increase their plants this way. The first step is to obtain seed—get it either from lowbush blueberries growing in your area or from the frozen "Wild" or Maine blueberries sold in many supermarkets. Give the seed a ninety-day rest if from fresh fruits, and start it in January, so that you'll have new plants in time for spring planting. Just fill a 3-inch-high flat with fine-ground sphagnum moss, sprinkle the seed evenly, and cover with a very thin layer of sphagnum moss. Keep the moss moist at a temperature of 70 to 80° F. until the seedlings emerge in 3 to 4 weeks. When they reach 2 to 3 inches, transfer them to peat pots filled with acid soil, which you can transplant to the garden after all danger of frost has passed.

Lowbush blueberries aren't the only unusual blueberry variety you can gather from the wild or try to grow in the home garden. Following are eight more types every bit as good in many respects and better in others. Save for noted exceptions, all are cultivated essentially like highbush blueberries.

Dryland

Sometimes called the *low huckleberry,* the dryland is native from Georgia and Alabama to Maine and westward to Michigan and Ohio. *Vaccinium pallidum* grows in dry, poor soils of hills and ridges and is very drought-resistant. Reaching 6 inches to 3 feet in height, it spreads much like the lowbush blueberry. Its large berries are sweet, tasty, light blue in color, and borne in dense clusters at the end of the previous year's wood, making them easy to pick. Drylands may someday be important in helping develop highbush blueberries that aren't so finicky about soil conditions.

Evergreen

Also known as the *coast* or *western blueberry, Vaccinium ovatum* grows 6 to 20 feet tall in its native woods along the Pacific coast from central California to British Columbia. Its berries are small, black, and of a strong flavor that makes them better for cooking in pies and other dishes than eating out of hand. Evergreen blueberries can be transplanted from the wild or propagated by cuttings 6 to 8 inches long. A handsome ornamental bush, this variety supplies florists with their leading foliage accessory, "evergreen huckleberry," the value of flowering branches taken from it worth more commercially than the fruit harvested from the bush. At one time its branches were picked so freely that the species was threatened with extinction.

Mountain

The "broadleafed huckleberry," as it is often called, is a native of the Sierra and Cascade mountains of Oregon and Washington and eastward to Wisconsin. *Vaccinium membranaceum* is an extremely drought-resistant plant that grows well in rocky soil. Its berries are among the largest and best flavored of all wild blueberries, black or maroon in color, and borne singly or in pairs, rather than in clusters like most blueberries. Mountain blueberries can be propagated from seed or cuttings and are best planted 3 × 5 feet apart.

Georgia

Vaccinium melanocarpum grows 3 to 7 feet tall and bears purple berries that are of fair to good quality. It needs moist soil and should be planted 3 × 5 feet apart.

Hairy

Another dwarf species that grows about 2 feet tall and spreads by underground rhizomes. Native to the mountains of North Carolina, *Vaccinium hirsutum* likes a sunny location and moist soil. The berries are covered with fine hairs, the fruit bluish black, sweet, and of good quality.

Southern Highbush

Vaccinium virgatum doesn't grow quite as tall as the northern highbush and its fruits aren't as good. Native to southern bogs and swamps, it has been used to hybridize highbush varieties good for the South. Should be planted 5 × 8 feet apart.

Canadian

Growing only 1 to 2 feet tall, *Vaccinium canadensis* is cultivated almost exactly like the lowbush blueberry. Its berries are more acid, ripen later, and the bush is very productive. Plant about 2 × 2 feet apart.

Dwarf

Vaccinium caespitosum is the smallest of the blueberries at 2 to 10 inches high and is well suited for rock gardens. This tiny plant, which grows from Labrador to Alaska and southward on mountain summits, has sweet blue-black berries that are a full ¼ inch in diameter.

Blueberry Recipes

Once harvested, blueberries should not be washed until just before being served. They may be stored a few days at room temperature away from sun and wind, but will last for weeks in the refrigerator if placed in dry, tightly covered glass jars.

Blueberries can be canned, dried, or frozen. The best way to freeze them is simply to spread them out on a cookie sheet without touching, freeze, and then pack them for storage in the freezer. The following recipes can be made with either frozen or fresh blueberries unless noted otherwise.

Blueberry Muffins

2 cups flour
½ cup sugar
½ teaspoon salt
2½ teaspoons baking powder
1 egg well beaten
¾ cup milk
1 cup blueberries
¼ cup butter, melted

Preheat oven to 400°. Sift flour and measure. Add sugar, salt, and baking powder, and sift again. Combine egg, milk, blueberries, and melted butter; add mixture to flour and stir just until blended. Bake in twelve ⅔-filled greased muffin tins for 20 to 25 minutes.

Blueberry Grunt

2 cups blueberries
½ cup sugar
1 cup water
1 cup sifted flour
2 teaspoons baking powder
¼ teaspoon salt
½ cup milk
1 pint heavy cream

Cook blueberries over low heat with sugar and water until soft. Mix and sift flour, baking powder, and salt, adding milk and stirring quickly to make a dough that will drop from the end of a spoon. Drop onto the boiling blueberries. Cook 10 minutes with the cover off and 10 minutes with cover on, and serve hot or cold with cream.

Blueberry Jam

4½ cups blueberries
Juice of 1 lemon
Grated rind of ½ lemon
7 cups sugar
1 cup liquid pectin

After crushing the berries, mix in lemon juice, grated rind, and sugar. Boil for three minutes, remove from flame, and stir in liquid pectin. Skim and pour into six sterile glasses, sealing them with paraffin when cold.

Blueberry Nut Bread

2 eggs
1 cup sugar
1 cup milk
¼ cup oil
3 cups flour
1 teaspoon salt
4 teaspoons baking powder
1 cup fresh blueberries
½ cup chopped nuts

Preheat oven to 350°. Beat eggs and add sugar, milk, and oil gradually. Sift flour with salt, and baking powder, and add to the liquid mixture. Stir just until blended. Fold in blueberries and nuts and pour into well-greased loaf pan. Bake for 50 to 55 minutes.

Blueberry Crisp

2 pints blueberries
¼ cup granulated sugar
2 tablespoons cornstarch
¼ teaspoon mace
Rind of 1 lemon, finely grated
2 tablespoons lemon juice
1 cup sifted flour
½ cup light brown sugar
Pinch of salt
6 tablespoons unsalted butter
½ cup walnuts, chopped
2 cups cream, whipped

Butter a 6- to 8-cup casserole dish and preheat oven to 400°. Wash and pick over blueberries. Pat dry, mix well with combined granulated sugar, cornstarch, mace, grated lemon rind, and juice, and spoon into prepared dish. Mix flour, light brown sugar, and salt in bowl. Cut butter into small bits and cut into flour mixture as if making pastry. Blend until crumbly, and mix in chopped walnuts. Sprinkle half of this streusel topping evenly over blueberries and cook for 10 minutes. Add rest of topping evenly and cook for another 10 to 15 minutes, until top is crisp and brown and juices are starting to bubble up. Serve hot with very cold cream or whipped cream.

Blueberry Pancakes

1 egg
¼ teaspoon salt
1 cup milk
2 teaspoons baking powder
1 cup flour
½ cup blueberries

Beat egg until light; add salt and milk. Sift baking powder with flour; stir it into first mixture. Beat until smooth. Add blueberries. Pour spoonfuls onto hot greased griddle.

Mom's Blueberry Pie

Plain pastry (see directions)
3 cups berries
1 cup sugar
2 tablespoons cornstarch
⅛ teaspoon salt
1 tablespoon butter
1 tablespoon lemon juice

Preheat oven to 425°. Make plain pastry by mixing and sifting 2 cups flour and ¾ teaspoon salt, then cutting in ⅔ cup shortening with a knife. Do not knead. Divide dough in two parts and roll out thin on a floured board. Line a pie pan with half the pastry, fill with the blueberry filling ingredients above (all mixed together thoroughly), cover with top crust and bake for ten minutes, reducing the oven to 325° F. for about 30 minutes to finish baking.

4

Growing Cranberries Without Getting Bogged Down

The cranberry sauce that the Pilgrims enjoyed at the first Thanksgiving dinner was no doubt called "fenberry sauce," because the cranberry was known as the fenberry in England at the time (it grows wild in fens or swampy places) and it wasn't until years later that it was given the name cranberry from the Dutch word for it, *kranbeere*. Nobody really knows where the name kranbeere came from, although one theory holds that early Dutch settlers here so named the berries because they were a favorite of cranes. The Indians, who used the cranberry as a food, for making a dye, and as a poultice for blood poisoning, called it *I-brimi,* the "bitter berry." Native cranberries that the Pilgrims found growing in New England bogs were much bigger than the English cranberries or the highly esteemed Scandinavian lingonberry, also true cranberries, and were appropriately given the scientific name *Vaccinium macrocarpon, macro* being the Greek for "large."

Cranberries, high in vitamin C, helped protect the first colonists from scurvy. American sailors from Nantucket and New Bedford later took casks of them to sea for the same purpose, much as the English used limes and lemons, and the tart berries were also mixed with dried venison for meat patties called pemmican, which were carried on hunting and exploring expeditions. Cranberry juice was, and still is, prescribed to help bladder infections, among a number of illnesses. The berry has traditionally been made into cranberry tarts, cranberry pudding, and, of course, the cranberry sauce that our Thanksgiving Day turkey wouldn't be complete without. It was even used by Massachusetts colonists to appease Charles II when in 1677 he angrily protested after the colonists coined their own "pine-tree shillings" without royal permission. Ten bushels of wild cranberries gathered from the bogs and sent to England helped assuage the king's anger.

Cranberries weren't cultivated to any extent until the early nineteenth century. It's said that one Henry Hall of Dennis, Cape Cod, was clearing a swamp patch when a sand slide buried his wild cranberries, and that he later adopted the practice of sanding cranberry bogs to increase size and yield when the vines responded with vigorous growth. The berries were among the first fruits to be canned, and cranberry jam processed by Underwood back in 1828 brought as much as $1.50 a can. Today New Jersey, Wisconsin, Minnesota (once called "the Cranberry State"), Oregon, and Washington all produce cranberries, but two thirds of America's 25,000-acre crop comes from the sandy bogs of Cape Cod, the greater part of it used to make cranberry sauce. There is, in fact, a museum called Cranberry World near Cape Cod, about a quarter of a mile from Plymouth Rock in Plymouth, Massachusetts.

While the true cranberry is difficult to grow, it is not impossible to raise in the home garden, as several writers claim, and gardeners can also try a number of excellent cranberry substitutes that are indeed very simple to care for. A third choice is to gather cranberries wild, the taste of uncultivated varieties being preferred by many gourmets. Directions for all three alternatives follow.

Gathering Wild Cranberries

True cranberries belong to the *Vaccinium* genus like blueberries and are also found in acid soils. However, the cranberry (except for the lingonberry and southern cranberry) only grows on *marshy* acid soils, especially peat bogs deep in the woods, and is never found in upland sites. Places where it grows have a low, flat topography, often with a sluggish stream close by, and vegetation that thrives near it includes sphagnum moss, the pitcher plant, the sundews, and even taller shrubs or trees like the white cedar, the tamarack, and the spruce. Several important species can be found in the wild, although the indiscriminate draining of land by developers has marked the end of many cranberry bogs.

Vaccinium oxycoccas. The northern cranberry, small cranberry, or marsh whortleberry. A native of northern America, this species has wiry creeping stems that root at intervals, evergreen, boxlike leaves less than a half-inch long, and pink flowers. Its small crimson berries, often spotted, are acid, and about ⅓ inch in diameter. They aren't as good as most true cranberries but will do if nothing else is available.

Vaccinium macrocarpon. The true American cranberry. Found from Newfoundland south to North Carolina, Michigan, and Minnesota. This prostrate plant looks much the same as the northern cranberry, but is coarser and bears bigger (twice the size) and finer berries, which are round, oblong, or pear-shaped, and vary from pink to very dark red, or may be mottled red and white. This cranberry has undergone much selective breeding and is the species cultivated by commercial growers, but isn't as flavorful as the next three species.

VACCINIUM QUADRIPETALUM. The Indian cranberry. A small-berried species native to the Northwest that the Indians taught settlers in the region to use.

VACCINIUM ERYTHROCARPUM. The southern cranberry. Indigenous in the mountains from Virginia to Georgia. Its small, dark-red berries are noted for their fine flavor, second only to the lingonberry among cranberries.

VACCINIUM VITIS-IDAEA. Known variously as the lingonberry, mountain cranberry, rock cranberry, cowberry, red whortleberry, or fox cranberry. Most famous under its Scandinavian name lingonberry (from the Swedish *lingon,* mountain cranberry), this small berry is used to make delicious lingonberry jam and lingonberry syrup for pancakes. Found wild in northern Europe and the eastern United States, it is a creeping dwarf evergreen shrub 4 to 9 inches high, with white or pink flowers grouped in clusters, shiny green leaves, and sparkling red berries. Another berry that doesn't need a bog to grow in, the lingonberry is found in mountainous moorland and pastures and, though a poor bearer, is the most delicious cranberry of all.

SUBSTITUTE CRANBERRIES AND HOW TO GROW THEM

A large number of berries are grown around the world as substitutes for true cranberries. None is quite as good as the real thing, but true cranberries are so difficult to grow that only the most dedicated gardeners attempt them, many others choosing these relatively care-free taste-alikes. Of all the principal substitutes, I have only listed sources for the first listed here, which is far better in all respects than any of the others.

VIBURNUM AMERICANUM (or TRILOBUM). The highbush cranberry, or the cranberry bush. This 8 to 12-foot-high shrub is the most widely grown cranberry substitute in America, and bushes are offered by several nurseries (including Southmeadow, Spring Hill, Field, and Ackerman). Sometimes it is botanically called *Viburnum Opulus* as well, and it may simply be a variety of that species, but in any event it has better fruit, and the leaves of *americanum* are smooth on the underside. Its leaves are maplelike, 3 to 5 inches long, and its white flowers are up to 4 inches wide. Interestingly, the larger marginal flowers on the cranberry-bush flower clusters are not males, but are completely sterile, their only function being to lure insects that will pollinate the fertile, less attractive flowers.

The highly ornamental cranberry bush, with its handsome red fall foliage, is exceptionally easy to grow and is a worthy addition to any shrub border; it will do well in sun or partial shade, and its scarlet berries are one of the few fruits that are rarely bothered by birds. Native to northern North America, or perhaps a variety of *Viburnum Opulus* brought here by English settlers, it is hardy from zone 4 northward and serves as an excellent substitute for cranberries in regions too cold for the latter. The cranberry bush thrives in a fairly moist, loamy soil, not needing an acid soil like true cranberries. Little cultivation is needed to grow it, and pruning consists merely of cutting some of the older shoots to the ground in the winter to encourage the growth of new shoots. The bush can be increased from seed—a good but time-consuming way to develop more flavorful types of berries—or from cuttings of ripe wood taken in early August.

Berries of the highbush cranberry are about as large as true cranberries and are very high in vitamin C, jell easily because they are high in pectin, and can be used in the same ways as the cranberry. Like cranberries they are not good for eating out of hand. The late Euell Gibbons recommended that they be used to make a healthful juice by simmering the berries in water, to which an orange's juice and rind have been added to mask the unpleasant odor, then straining and reheating the brew to boiling.

VIBURNUM OPULUS. The cranberry tree, the snowball, the guelderberry, the guelder rose. Similar in almost all respects to the viburnum cranberry bush above, the guelderberry is another of the several viburnum species used as a cranberry substitute. It is found in the wild in the same damp woodlands and thickets and is cultivated the same way. The main differences are that its leaves are longer-lobed, hairy instead of smooth on the underside, and its fruit is more bitter. The shrub is useful in cities, as it is highly resistant to smoke pollution. Hardy from zone 2 southward.

VIBURNUM OPULUS LUTEUM. A yellow-berried variety of the above, which, while it isn't the right color for a cranberry substitute, is sometimes used to make wine. From zone 2 southward.

VIBURNUM ALNIFOLIUM. Hobblebush, hobbleberry, witch-hobble, wayfaring tree. A 6 to 10-foot-high spreading shrub, native to mountainous, eastern areas, with fruit that turns from red to purplish-black when fully ripe. It is called hobblebush because its lower branches root at the node, lying flat on the ground, and sometimes trip or hobble unsuspecting passersby. From zone 2 southward.

VIBURNUM LENTAGO. Nannyberry, sheepberry, wild raisin. A treelike type of viburnum that can grow 30 feet tall, has a great abundance of flowers and juicy bluish-black fruits with a slight bloom. From zone 2 southward.

VIBURNUM PRUNIFOLIUM. Stagberry, stagbush, black haw. A shrub or small tree 10 to 15 feet high, with small, plum-treelike leaves and bluish-black fruits with a slight bloom. Will grow in drier soils than most viburnums. From zone 3 southward.

VIBURNUM NUDUM. Possum haw. A native American shrub growing 6 to 12 feet tall, the twigs a little scurfy. Likes a somewhat acid soil and bears sweet, blue-black berries. From zone 3 southward.

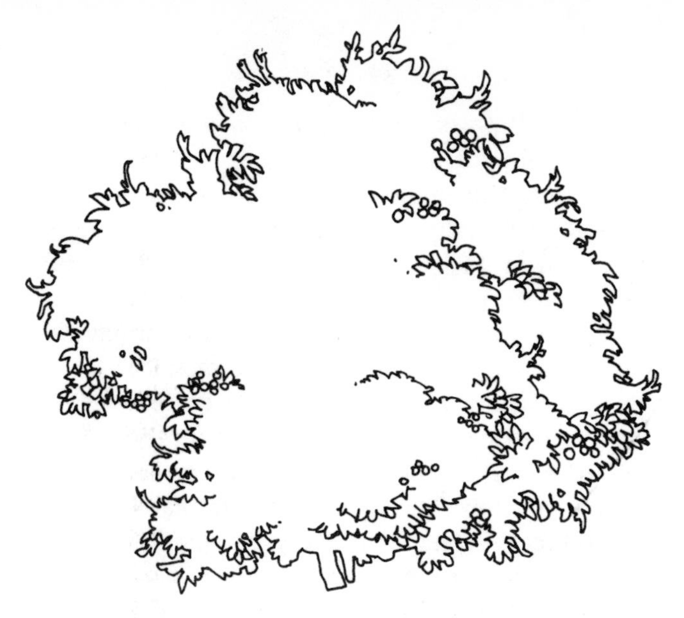

Cranberry bush, an 8 to 12-foot-high shrub is the most widely grown cranberry substitute.

VIBURNUM CASSINOIDES. Withe-rod, Appalachian tea. An attractive, 8 to 12-foot-high, well-rounded shrub that bears sweet bluish-black berries and prefers a slightly acid soil. From zone 3 southward.

ASTROLOMA HUMIFUSUM. The Tasmanian trailing cranberry. A handsome evergreen shrub that grows to a height of about one foot and has solitary scarlet flowers and red berries. The prostrate shrub, which is many-branched, thrives in an equal mixture of sand, loam, and peat with thorough drainage if grown in climates similar to its natural home or in a cool greenhouse. Unfortunately, it is offered by no nurseries in the U.S. as far as I can find, so an Australian source would have to be located. It is propagated by young cuttings, which root readily in sandy soil under a bell glass. The name *Astroloma,* incidentally, is from *astron,* a star, and *loma,* a fungi, in reference to the bearded limb of the coralla.

LISSANTHE SAPIDA. The Australian cranberry. As its name *sapida* indicates, this Australian berry is "savory." A pretty evergreen shrub that grows to four feet high in its native western Australia and Tasmania has spikes of white flowers tipped with green and red berries. Unfortunately, like the above plant, it is not offered by U.S. nurseries.

GROWING TRUE CRANBERRIES

If substitutes aren't your thing, and you decide to experiment growing real cranberries, there are several alternatives open. The easiest is to plant either the previously mentioned lingonberry (*Vaccinium Vitis-idaea*) or the southern cranberry (*Vaccinium erythrocarpum*), neither of which need to be grown in peat bogs. Both of these berries do need an acid soil of about pH4.5 per cent and if the spot you pick for them isn't suitable you can make the soil acid in the same way as described for blueberries (see Index). The lingonberry makes an especially good ground cover with its shining green leaves and will do well in half shade, while both plants are often grown in peat pockets made in rock gardens. Neither is a great yielder, so you'll need a lot of plants if they're grown for the berries, but both have a more highly esteemed flavor than the bigger-fruited true American species. Care for them in the same way as for the American cranberry, following.

Raising the large-fruited American cranberry of the commercial growers (*Vaccinium macrocarpon*) is a tricky proposition at best and isn't recommended unless you have the ideal location: that is, an acid peat bog with a clay subbase that will maintain a steady supply of acid water in the upper layer of muck, *and* a relatively sluggish stream flowing nearby. Growers with such ideal natural conditions begin operations by building a small dam with gates at the start and flooding the area to kill all vegetation. Then clear, moist sand is added atop the peat bog to a depth of 2 to 3 inches and cranberry plant cuttings are inserted through the sand into the peaty soil below in rows 12 to 18 inches apart, with two cuttings put at 8-inch intervals in each row. Early Blacks, Howes, Wilcox, Beckwith, and Stevens are excellent varieties to plant and will bear fruit from cuttings in 2 to 3 years (it will take a year longer from seeds stratified in the fall and the seeds will not come true to form). The bogs are flooded (submerging the plants) from December to May every year to control insects and diseases and to prevent frost damage. It is essential to keep the plantation weeded until the plants become established, but after that, weeds are rarely a major problem. Bogs are usually resanded every two years. They are fertilized every year with a formula of 75 pounds sodium nitrate, 75 pounds dried blood, 300 pounds rock phosphate, and 50 pounds of sulfate of potash at the rate of 500 pounds per acre. Cranberries raised in this manner yield from 100 to 300 bushels an acre. Plants can be increased by division, by layering of stems in summer, and by cuttings.

The 999,999 gardeners in a million who don't have a bog and a small, sluggish stream nearby and still want to grow large-fruited cranberries might possibly simulate the necessary growing conditions. One method would be to dig out a low-lying location to a depth of about two feet and fill it half with stones and then half with acid peat and leaf mold as you would for a blueberry patch, making sure that the soil pH is about 5.0. After planting, either keep the plantation flooded from December to May with a garden hose as needed, or keep the plants watered during the growing season with a sprinkler system and mulch them over the winter with peat moss. Aside from this, the plants should simply be kept free of weeds. The main trouble with sprinkler watering is that it doesn't destroy the insect enemies of the cranberry—especially the cranberry leafhopper, a pest that transmits a virus called cranberry false blossom, which causes abnormal blossoms that don't produce fruit and has wiped out commercial crops in the past. Using somewhat resistant varieties like Wilcox and Stevens, which the leafhopper doesn't like to eat, helps here, however, and in a small home planting, infested or diseased plants can be pulled out to prevent the disease from spreading.

Harvesting Cranberries

Cranberries were once called "bounceberries" because their ability to bounce is a test of their excellence. In early times, harvested berries were rolled down a series of steps, the good firm ones bouncing like little rubber balls to the bottom and the soft damaged ones remaining on the steps. Today's grading machines work on the same principle. Whether cranberries are harvested from wild bogs or from the home garden, it's best to use a wooden cranberry scoop that combs the berries from the vines. The berries should be harvested (or flooded) before the thermometer reads much below 27° F., for colder temperatures will quite likely ruin them. They should be used soon after harvesting, or frozen for later use, since they do not keep well, being subject to various rots after picking.

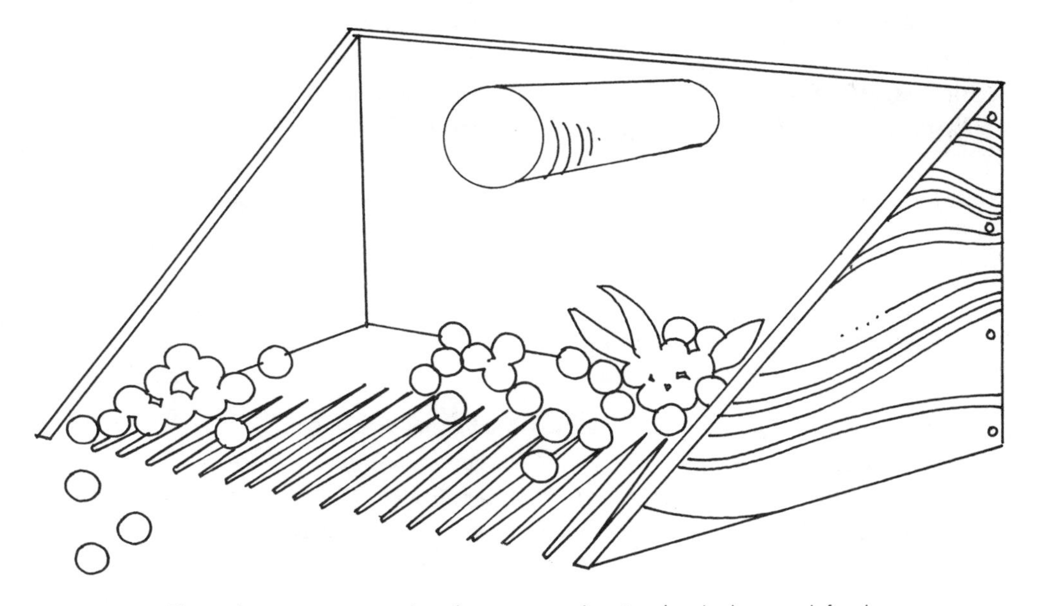

The rake or scoop used to harvest cranberries (and also used for low-bush blueberries).

Cranberry Recipes

All of the following recipes can be made with cranberry substitutes as well as true cranberries, but for the best results the berry called for should be used.

Cranberry Soup

½ cup cranberry sauce
1 sliced onion
1 cup chopped cabbage
6 cups cold water
½ cup beet juice
1 cup chopped, boiled beets
1 teaspoon salt
1 tablespoon sugar
¾ cup sour cream
3 hard-boiled eggs

Mix cranberry sauce, onion, cabbage, and 6 cups cold water. Put in saucepan and boil 20 minutes, stirring occasionally. Add beet juice, beets, salt, and sugar. Put 2 tablespoons sour cream in each plate. Pour in soup. Garnish each plate with ½ hard-boiled egg. May also be served cold.

Cranberry Turkey Stuffing

1 cup fresh stewed cranberries
¼ cup sugar
¼ cup chopped celery
2 tablespoons chopped parsley
4 tablespoons butter
4 cups bread crumbs
½ teaspoon sweet marjoram
1 teaspoon salt

Combine cranberries and sugar. Cook celery and parsley in butter until celery is tender. Blend all ingredients together.

Cranberry Orange Bread

2 cups flour
1½ teaspoons baking powder
½ teaspoon baking soda
½ teaspoon salt
1 cup sugar
2 tablespoons salad oil
¾ cup orange juice
1 egg, beaten
1 cup sliced raw cranberries
1 cup chopped nuts

Preheat oven to 350°. Sift all dry ingredients. Mix in oil, juice, and egg thoroughly. Fold in cranberries and nuts. Pour into oiled loaf pan. Bake for 1 hour.

Easy Cranberry Sauce

Boil 1 cup of sugar and 1 cup of water together for five minutes. Add 2 cups of cleaned cranberries, letting them cook five minutes or until the skins break. Cool; then refrigerate until thick.

Old-fashioned Cranberry Catsup

Pour 2 pounds of cranberries and 1 cup each of vinegar and water into an iron kettle. Boil until the berries are pulped, rub them through a sieve, and return them to the kettle with 2 cups of dark-brown molasses and ½ teaspoon each of ground nutmeg, cinnamon, allspice, cloves, and salt. Let the catsup boil five minutes. Makes one quart.

Candied Cranberries

½ cup firm, cleaned cranberries
½ cup sugar
½ cup water

Prick each cranberry with a needle. Boil sugar and water until syrup spins a thread (234°). Add the berries and keep cooking until the syrup forms a hard ball in cold water (250°). Take berries from syrup and let stand on wax paper till dry. Roll in granulated sugar.

Simple Cranberry Sherbet

4 cups fresh cranberries
2½ cups water
2 cups sugar
1 teaspoon gelatin
½ cup cold water
Juice of 1 lemon
Juice of 1 orange

Cook berries in water until they stop popping. Strain. Add sugar; cook until sugar dissolves. Add gelatin that has been softened in cold water. Cool. Stir in fruit juices; pour into ice cube tray. Stir twice during freezing.

5

Heavenly Huckleberries

No fruit is better known by name in America than huckleberries—due in large part to Mark Twain dubbing his most memorable character Huckleberry Finn—but few Americans have ever tasted the native delicacy. This wasn't always so, for the huckleberry has a long history as an edible fruit. American Indians valued the berry highly, relishing one feast dish made of "dog's flesh, boiled in bear's grease with huckleberries." Cherokee women encased huckleberries in corn-meal dough, deep-frying them in bear fat, and similar dishes featuring them are still served at the annual feast of Cherokee foods held in Cherokee, North Carolina.

The first American settlers noticed the wild huckleberry, comparing it with the English bilberry and first calling it a *hurtleberry* or *hirtleberry,* from which its present name derives. One colonist noted in 1607 that the soil near the mouth of the James River in Virginia "naturally yields mulberry trees, cherry trees . . . gooseberyes, strawberyes, hurtleberrys" and other small fruits. "Huckleberrying" or "ahuckling" early became a popular American summer pastime, with outings held annually to gather the berries from the translucent purple and crimson "huckleberry bogs," "huckleberry cuts," and "huckleberry slashes," which often harbored wolves, foxes, and many other wild animals. Huckleberry pie and huckleberry pudding were great native favorites from the beginning, and there was even a huckleberry johnnycake made half of meal and half of fresh ripe huckleberries.

Huckleberries were so little, plentiful, and common a fruit that "a huckleberry" became early nineteenth-century slang for a small amount or a person of no consequence, both of these expressions probably inspiring Mark Twain to name his hero Huckleberry Finn. The berry was also immortalized in the colloquial phrase "as thick as huckleberries," very thick, and "to get the huckleberry," to be laughed at or ridiculed, a predecessor of sorts of the raspberry (razz) or Bronx cheer. "To be a huckleberry to someone's persimmon" meant, in nineteenth-century frontier vernacular, to be nothing in comparison with something else, and "to be a huckleberry over someone's persimmon" meant to outrank someone. The huckleberry's abundance and commonness was reflected in many poems and stories, including Whittier's lines: "Dread Olympus at his will/Became a huckleberry hill."

The huckleberry is not a bilberry—its fruit grows in clusters, for one difference—and it is definitely not a blueberry, though many writers have used the names interchangeably. Though they belong to the same heath family (Ericaceae) as blueberries, huckleberries are smaller and seedier. They contain 10 seedlike nutlets, comparatively large, bony seeds that are quite noticeable (crackling between the teeth as you eat them), while the blueberry has 60 or 70 very small soft seeds that aren't noticeable at all. Another way to distinguish the two plants is by the yellowish dots on the underside of huckleberry leaves; blueberry leaves have none.

Huckleberries, which are not a true berry but a drupe fruit, belong to the *Gaylussacia* genus, which was so named in honor of French chemist Joseph Louis Gay-Lussac (1778–1850). One species of the genus, *Gaylussacia brachycera,* or the wild box huckleberry, holds the world record as the plant covering the greatest area from a single clonal growth—a colony that began an estimated 13,000 years ago and covers an area of 100 acres was found in 1920 near the Juniata River in Pennsylvania. Not nearly so big as the blueberry, though some people prefer its "wild" flavor, huckleberries grow up to ¾ inch in diameter and aren't edible until they are jet black and very soft. Unlike blueberries, they are rarely cultivated in the home garden and little, if anything, has been done to improve them by plant breeding over the years. Nevertheless, the tasty berries are worthwhile gathering for pies and preserves. They can even be grown in the home garden if you have enough room and are energetic enough to collect plants from the wild or propagate them from cuttings—for no nurseries, to my knowledge, offer true huckleberries. Here are the species to look for, or avoid, when gathering berries or looking for bushes to transplant.

GAYLUSSACIA BACCATA. A deciduous shrub 1 to 3 feet tall that is called the black huckleberry and also the highbush huckleberry, this plant bears the sweetest huckleberries, the shiny black berries growing about ⅜ inch in diameter and ripening in June. The flowers are pink or pale red and the young growth on the plant is sticky and resinous, leading it to be classified sometimes as *G. resinosa.* Leaves are oblongish, up to 3 inches long, and yellowish-greenish above, dotted underneath. Unlike most huckleberries, this species is at home in dry, sandy, or rocky soil. It is found throughout eastern North America.

GAYLUSSACIA FRONDOSA. A taller plant that grows from 3 to 6 feet high, the dangleberry or tangleberry, as this deciduous huckleberry is commonly called, bears bigger berries than the black huckleberry. The fruit is ½ inch in diameter and is dark blue with a whitish bloom, sweetish, and pleasant-tasting. Dangleberry bushes prefer moist, peaty soil and are usually found in woods and thickets. The flowers are greenish-purple and the leaves on the spreading shrub are somewhat elliptic, up to 2½ inches long, green above and pale green and hairy underneath. This species grows wild from New Hampshire to Florida and Louisiana.

GAYLUSSACIA DUMOSA. The bush huckleberry, also called the gopher-berry, and the dwarf huckleberry, only grows to about 1½ feet tall and spreads by means of underground stems. The deciduous bush has white flowers and small black berries that taste rather insipid but are edible. Its deciduous leaves are oblongish, nearly stalkless, and up to 2 inches long, resinous underneath. Found from Newfoundland to Florida, mostly in sandy bogs.

GAYLUSSACIA URSINA. The bear huckleberry. A deciduous shrub that grows 2 to 6 feet tall, has white or red flowers and small black berries. Found from New Hampshire to Florida.

GAYLUSSACIA BRACHYCERA. The box huckleberry. Named for its box-shaped leaves, this nearly prostrate evergreen shrub rarely grows over 1½ feet high and is valued as a ground cover. It has white or pink flowers and blue fruit that isn't of much value. Also called the juniperberry, it is found from Delaware to Tennessee.

SOLANUM NIGRUM. The garden huckleberry. Developed by Luther Burbank from, of all things, the highly poisonous plant black or deadly nightshade. The garden huckleberry, also called the wonderberry and sunberry, is, of course, not a true huckleberry, being more closely related to the tomato than any other berry. It is simply an improved form of deadly nightshade that Burbank developed, though nursery catalogs never tell you its origins. You should be aware of them nevertheless and *not* collect the garden huckleberry from wild forms. Seed is available for this novelty from Burgess and it is easily grown as an annual in any slightly acid garden soil. Start the seeds indoors and set outside after all danger of frost has passed, feeding and watering often throughout the season. The blue-black berries from the 1 to 2-foot-tall plant are not good eaten out of hand, but make excellent preserves, a delicious garden huckleberry pie that tastes like green tomato pie, and are good for freezing and canning. *Remember that the juice of the plant's wilted leaves is dangerously poisonous.*

Raising true huckleberries is not much different from growing blueberries (see Index), except that huckleberries like partial shade while blueberries prefer full sun. Both are acid-loving plants that are difficult to grow in a soil with a high lime content. If you have room for a few bushes—and they are easy to find space for, being ornamental and preferring shady spots as they do—collect several wild plants from one of the above species and transplant as you would blueberry bushes. Or propagate new bushes from cuttings taken from tips of the current year's growth in August, inserting these in a cold frame filled with sandy peat moss until they are well rooted. After that, plant and care for them just as you would for blueberries, which taste tame in comparison.

Huckleberry Recipes

Huckleberry pies, puddings, tarts, preserves, and wine were recipes quite familiar to early Americans. Whether you collect the berries from the wild or grow your own, try at least one of the following:

Garden Huckleberry Pie*

2 tablespoons tapioca
1 cup plus 2 tablespoons granulated sugar
1/8 teaspoon salt
1 quart stemmed garden huckleberries
1 tablespoon lemon juice
Pie pastry
1 tablespoon butter
Pinch of nutmeg

Preheat oven to 350°. Mix tapioca, sugar, and salt, and sprinkle over garden huckleberries, adding lemon juice. Pour this filling into the pie shell, dot with butter and nutmeg, and cover with top pie crust, baking until berries are tender, about 45 minutes.

* Use only garden huckleberries (*Solanum nigrum*).

Huckleberry Pie

Use the same recipe as for blueberry pie (see Index), substituting an equal amount of true huckleberries for blueberries.

Almost any blueberry recipe can be made with huckleberries.

Huckleberry Pudding

¼ cup shortening
¾ cup sugar
1 cup sifted flour
¼ teaspoon salt
1 teaspoon baking powder
¼ cup milk
2 egg whites, beaten
1 teaspoon vanilla
3 cups true huckleberries

Preheat oven to 350°. Cream the shortening and stir in ½ cup sugar gradually. Sift together the dry ingredients, except sugar, and add alternately with the milk to first mixture. Fold in the stiffly beaten egg whites. Add vanilla. Mix huckleberries with the remaining ¼ cup sugar. Place in bottom of greased pudding dish. Pour batter over them and cover mold tightly. Bake for 1¼ hours.

6

Magic Elderberries

He who cultivates the elderberry," instructs an old proverb, "will die in his own bed" (not from the elderberries, I hope!). Ever since Hippocrates recommended them, amazing properties have been attributed to the wine, pies, preserves, and medicines made from *Sambucus nigra*. Wrote seventeenth-century English herbalist John Evelyn: "If the medicinal properties of the [elderberry] leaves, bark, berries, etc., were thoroughly known I cannot tell what our countrymen could ail for what he would not find a remedy, from every hedge, either for sickness or wounds." With endorsements like that it's a wonder that more people don't cultivate the "magic berry" today.

The elderberry or elder tree is, if nothing else, a real conversation piece, the subject of superstitions or beliefs that go far back in time. According to one story, Christ's cross was made from elder wood, and a medieval tradition held that Judas hung himself from an elder tree. Believing he had sinned and betrayed the innocent blood, Judas returned the thirty pieces of silver he had been paid for his treachery and went out and hanged himself. The elder he chose became known as a "Judas tree," its flower a "Judas blossom," and the edible fungus cup that sometimes grows on the tree has since been called a "Judas ear."

The Germans believed that men should doff their hats in the presence of an elder tree and in Denmark the tree was thought to be under the protection of a jealous "Elder-mother" who forbade anyone to gather the tree's flowers without her permission and would not let its wood be used to make any household furniture. A child sleeping in an elderwood cradle, the Danes believed, would surely be strangled by the Elder-mother.

Nevertheless, the wood of old elders, which is white, hard, and close-grained (Shakespeare wrote of a character's "heart of elder"), has been used to make a number of articles over the years, including skewers, combs, and shoemaker's pegs, and from early times young elder twigs, their pith removed, have been employed in making whistles, pop guns, and other toys.

In the days when sugar maples were tapped with an auger, an "elder quill" (a small 3-inch piece of elder with the pith bored out of it) was inserted in the incision to conduct the sap into the trough below. For that matter, practically every part of the elder has been used by man. The leaves and bark emit a sickly odor believed to be effective as an insect repellent, and the inner bark was once administered as a purgative. Elderberry leaves were employed to impart a green color to fat and oil, and the elder's flowers contain a volatile oil that serves for the distillation of elder-flower water, which is used in candy, perfumes, and lotions. Early American colonists trusted in the medical powers of both elder-bark tea and an elder-blow tea made from elderberry blossoms. This last was used well into the twentieth century (and maybe still is) as a laxative, a diuretic, a febrifuge to reduce fever, and even as a poultice when "combined with a biscuit."

It's been said that almost all of the elder is edible, that there is something to eat from the tree in every season. The berries, of course, are used to make elderberry wine, elderberry jelly (which needs no pectin added), and delicious elderberry pie, an old country favorite, not to mention spiced elderberries, an elderberry sauce that is used like cranberry sauce, and an elderberry syrup very high in iron and vitamin C (only black currants and rose hips are higher) that helps ward off colds. But elderberry blossoms can be used to make wine, too (a wine called elder-blow), and elderberry blossom fritters are a gourmet's delight that taste something like waffles. Finally, the young leaves of the plant can be scalded in boiling water and added to a salad, while green elderberry stalks (after paring away all the woody outer casing) can be used in the same way as artichoke stalks or endive stumps. Such a versatile, attractive, all-purpose plant, which is a great favorite of birds and has a delightful odor when in bloom that Emerson admired, certainly deserves a place somewhere in the small fruit garden, along a fence row, or in the shrub border. Especially since it is virtually care-free.

The Best Elderberry Varieties

Here is a fruit that you as a gardener can improve by selection and cultivation. Comparatively little work has been done breeding elderberries over the years, even though there are commercial elderberry growers producing the berries for jams and jellies. Only a relative handful of varieties are available, all of them, however, much superior to the wild types in size, quantity of berries, and juiciness. Most nurseries selling elderberries claim that the plant is not self-fruitful and that two must be planted for cross-pollination; in fact, several nurseries will only sell the bushes in pairs. The truth is that some varieties, including the popular Adams, do need cross-pollination unless there are wild elderberries nearby and it is necessary to plant both green-stemmed and red-stemmed clones, or types.

I know of only seven elderberry varieties that can be obtained from

A full-grown elderberry bush. Elderberry bushes often grow 20 feet high and more.

nurseries. Adams, one of the first varieties of elderberry offered, was selected from the wild in New York in the early 1920s. A 5 to 6-foot-tall bush—a few feet smaller than most wild elders and much less sprawling in habit—it bears huge clusters of purple-black berries about ¼ inch in size. It is offered by Kelly Brothers, Zilke Brothers, and Burgess. Most other varieties are slight improvements on Adams. Kent, a very productive bush, and Johns Improved, which ripens later than Kent, are offered by Henry Field's, while York and Nova, two new, highly rated varieties, are carried by Zilke Brothers. The New York State Fruit Testing Cooperative Association has available Ezyoff, a variety whose berries can be stripped from their stems more readily than most, and New York "21," the very latest development, which is a cross between Adams and Ezyoff. Plentiful seeds are the major objection to elderberries, but no one has yet perfected a relatively seedless fruit.

Gathering Wild Elderberries

If you want to pick wild elderberries or find some excellent bushes suitable for transplanting to the garden, there are a number of varieties to choose from if herbicides and land developers don't get to them before you do. Wild trees, larger than home-grown types, usually grow in the open, often in elder swamps, alongside streams and in other moist places. The *Sambucus* genus belongs to the Caprifoliacae family, which includes the genus *Viburnum*. There are some twenty species, all deciduous with pithy branches, opposite, compound leaves, all toothed, and small flowers in terminal, many-branched, umbrellalike clusters. The fruit is berrylike, with 3 to

5 one-seeded nutlets. Following are the commonest edible North American varieties. Not included here is the European elder, *Sambucus nigra,* which grows up to 30 feet tall:

- American or Sweet Elderberry (*Sambucus canadensis*). This 6 to 12-foot-tall species bears purplish-black fruit about ⅙ inch in diameter in late June–July. The fruit is larger, sweeter, and more abundant than most wild elderberries, and the bush stands smoke better than most shrubs.
- Blue-berried or Western Elderberry (*Sambucus caerulea*). A shrub that can grow from 10 to 40 feet high, has yellowish-white flowers like the American elderberry above, and bears larger, bluish-black berries at about the same time of year.
- Red-berried Elderberry (*Sambucus racemosa*). An attractive 8 to 10-foot shrub with yellowish-white flowers that is probably the handsomest species and bears scarlet berries from late April to June and sometimes July, depending on the variety.
- California Elderberry (*Sambucus callicarpa*). Growing only to about 6 feet tall, this California native bears small scarlet berries.
- * Blue Elderberry (*Sambucus glauca*). A 15 to 30-foot shrub with yellow-white flowers and sweet bluish-black berries that ripen in late June, early July.
- Golden Elderberry (*Sambucus pubens zanthocarpa*). A shrub rarely over 10 feet high that bears golden-yellow berries.
- White Elderberry (*Sambucus pubens leucocarpa*). Similar to the above, but bearing white berries.

* Fruit is covered with a delicate powdery blue "bloom" or "farina" (the cause and use of which is unknown) which gives it a distinctive hue.

When, Where, And How To Plant

Elderberries, which can be grown in all but the very coldest areas of North America (zone 1 and parts of zone 2), prefer a sunny location in the garden, although they can take partial shade. They should be set out in early spring or late fall and thrive best in moist, loamy, well-drained places, but they will grow in a wide range of soils. Just don't plant them where the soil dries out quickly or where their feet will always be wet. Because the shrubs are inclined to sprawl, grow tall, and spread by means of suckers from the roots, they are usually planted in the shrub border rather than the small fruit garden. If they are grown in the cultivated fruit garden, be sure to plant them where they do not shade other sun-loving berries, remembering that they can reach heights of up to 40 feet, and plant them at least 5 feet apart. To stop them from spreading where they aren't wanted, keep an eye out for the suckers they send out, tearing them from the ground. Named varieties are less aggressive in their spreading habits than the wild types.

Space Savers

Elderberries are useful for making orchard screen-fences in exposed situations, as a shelter for other shrubs, and for making hedges that provide a useful fruit as well as privacy. To keep them close and compact in growth for these purposes they need only be clipped two or three times a year.

Care Of Elderberries

Fertilizing, Watering, and Weeding

The elderberry is an easy plant to grow, but several pointers should be kept in mind for bigger and better crops. It is important to keep the soil moist around elderberries at all times—especially the first year they are planted—and the best way to do this is to apply a 2 to 4-inch mulch of hay or leaves, which will keep down weeds as well.

Elderberries will do just fine without fertilizing, but for more and bigger berries, commercial growers regularly feed them. The reason for this is obvious if one studies the bearing habits of the plant. The elderberry bears fruit on shoots for 3 to 4 years, but these shoots bear less every year before they eventually die. It is on new shoots formed every spring that the plant bears most abundantly. Therefore, commercial growers prune the shrubs back before new growth starts in the spring and apply nitrogenous fertilizer, which encourages the new growth that will bear the most flowers and berries.

Pruning Them

Aside from the hints above about pruning elderberries for a space-saving hedge and to encourage the production of young fruit-bearing shoots, the shrubs need only be pruned to cut out dead wood, three-year-old wood, and crisscrossing branches. The bush can also be cut back to prevent crowding and keep it at any height or width you choose. To keep the bush compact as a specimen plant, just clip off any shoots growing too far away from the plant and don't let suckers form from the roots.

Insect Pests and Diseases

The major culprit here is the Elder Borer or Spindle Worm, which causes the ends of elderberry shoots to blacken and fall off. This can be controlled by cutting out the dead wood and burning it, or disposing of it in another sanitary manner. Voracious birds, which love elderberries and often strip a bush of fruit, can be controlled by netting the plants if they are kept compact enough (see Appendix I, For the Birds). Otherwise, elderberries seem freer from pests than any berry.

Propagation

Elderberries are easily increased by hardwood cuttings, suckers, or seed, and often birds will save you the trouble of planting by dropping seeds themselves.

Elderberry hardwood cuttings are usually made in late autumn or early spring. Just take dormant wood 9 to 18 inches long from the bushes, making certain there are three sets of opposite buds. Bury each cutting where you want the elderberry to grow, sticking it in the ground up to its pair of top buds, and keeping it watered during the growing season.

Suckers (which were once called "thieves" because they rob the parent plant of nutrition) are simply dug up from their elderberry mother plant to make new bushes.

Elderberries can also be increased by root division. Just dig up a clump in early spring, divide it into two or more sections, and replant the sections for as many new bushes as you want.

Harvesting Elderberry Fruit And Flowers

As with currants, the little ¼ to ½-inch elderberries should be picked by the cluster rather than individually. Once you have a basket of these, just sit down in a comfortable spot and strip the berries from the clusters. Be careful not to strip off stem pieces with the berries, for this can make the flavor of elderberry wine, jam, and pie too bitter. When picking berries for jelly, be sure that they are only half ripe or that there are a number of green ones so that no pectin needs to be added to the recipe.

Elderberry blossoms are harvested for elderberry fritters just before the fruit forms, leaving a bit of stem on them. Infinite patience is required in gathering them, but the result is well worth the effort.

Elderberry Recipes

Elderberry Blossom Fritters

This European delicacy is made by holding each elderberry blossom by its stem and dipping it into a thin pancake or waffle batter, then frying in deep oil until golden brown. The fritters can be rolled in powdered sugar when done.

Tasty Spiced Elderberries

1 tablespoon whole cloves
1 tablespoon whole allspice
1 stick cinnamon
3 pounds sugar
1 pint cider vinegar, diluted with water
5 pounds of elderberries

After tying spices in a cheesecloth bag, heat sugar, vinegar, and spices to boiling and cool. Add berries and simmer until tender, cooling and letting stand overnight before removing spice bag.

Elderberry Apple Pie

Use the recipe for your favorite apple pie and add half elderberries instead of all apples. This will make the elderberries seem less seedy. If the elderberries are first dipped in honey and stored in glass or plastic containers at room temperature until they become slightly "winy" in taste, they make even better pies.

Elderberry Gourilos

Pare all the woody casing away from young elderberry stalks until they are green and prepare in the same way as artichoke stalks. That is, cut the stalks into 2-inch-long sticks and blanch in water flavored with lemon. Once the sticks are blanched they can be simmered in butter or cream in a covered pot or sautéed in butter.

Elderberry Wine

Put only enough water with the berries to keep them from burning and cook a few minutes, just long enough to scald thoroughly. Strain through a cheesecloth; add 8 cups of sugar to 10 cups of berry juice. Set away in a cool place to ferment, and skim daily until clear. When bubbles cease to rise on top of the liquid, it is ready to bottle. Use only white sugar and you'll have some of the richest-flavored wine you ever tasted. Yeast is apt to sour the wine and it is a mistake to hurry the fermenting process. The addition of other liquor spoils the fine, natural flavor.

7

Backyard Blackberrying

One of the oldest fruits known, blackberries were doubtless gathered and eaten by the earliest inhabitants of earth, who were fruit-eating gatherers rather than omniverous, according to the latest anthopological theory. These plump, delicious bramble fruits, a member of the vast rose family like the raspberry, are native to the temperate regions of Asia, North America, and Europe as well as South Africa. They are mentioned in the writings of Aeschylus and Hippocrates 500 years before Christ, and the ancient Greeks not only enjoyed them but thought that they prevented gout and other ailments, including diseases of the mouth and throat. The English herbalist Culpeper prescribed them for snakebite, kidney stones, quinsies (tonsillitis), and even recommended the leaves boiled with lye as a dye to "maketh the hair black."

In England blackberries are the most common fruit growing in the wild and proverbially have come to represent what is plentiful because they outyield all other bramble fruits—one plant can yield up to five gallons of berries. "Plentiful as blackberries" comes to us from Shakespeare (though what he actually wrote was: "If reason were as plenty as blackberries, I would give no more reason on compulsion").

The English call a cold early May when blackberries are in bloom a "blackberry winter" and in America a "blackberry summer" is a period of fine weather in late September and early October. Americans have probably prized the blackberry more than any other people. It is the subject of one of the best still lifes by an American artist, Raphaelle Peale's tiny "Blackberries" showing a porcelain dish of blackberries, leaves attached, that seem so vividly real that they transcend reality. In early America blackberries were eaten fresh with cream or wine, made into syrups, jams, pies, cobblers, cordials, wine, cold summer drinks called "fruit waters," and flummeries—molded desserts made of gelatin, milk, blackberry juice, and sugar. A black dye was made from blackberry roots. Blackberry tea was supposed to be a sure cure for dysentery, the plant's bark, root, leaves, and fruit being astrigent and high in tannins. In fact, temporary truces were declared during the Civil War so that both Union and Confederate soldiers plagued with dysentery could go "blackberrying."

Blackberries were found by early Americans to be a refreshing, nourishing fruit (they contain 85 per cent water and 10 per cent carbohydrates, in addition to a high content of mineral salts, vitamin B, vitamin A, and calcium, which is present in them more than in any other fruit). Although most early agricultural papers mainly gave advice on how to kill the abundant wild blackberries in America to clear land for farming, Americans cultivated the fruit long before it was domesticated in Europe and brought under subjugation at least seven wild types of the fruit in less than one hundred years, developing many excellent varieties that are still grown today. Walt Whitman best expressed the national fondness for blackberries when he wrote that the berries "adorned Heaven."

Going Blackberryin'

The blackberry derives from various species of the 200-odd members of the *Rubus* genus. Unlike the raspberry, another Rubus family member, and like its cousin the dewberry, its mass of fleshy druplets or seeds adheres to its fleshy receptacle, which is part of the berry and is eaten with it. To put it another way: hollow, thimblelike raspberries are pulled away from their core or receptacle when picked and their cores aren't eaten, while blackberries are picked core and all and their cores are eaten. Since the hard, green cores aren't particularly tasty, the best blackberries are plump, soft, and small-cored. If you go blackberrying for wild berries, avoid hard, smaller berries and pick only the large ones. You can also look for superior wild bushes (those that are prolific produce large berries, etc.) and transplant them to the home garden, but although this has the advantage of assuring winter-hardy plants for your locality, the wild plants may be infected with some disease that isn't apparent and infect other blackberry plants in your garden. Anyone who lives in the country, however, can usually find enough wild bushes out in the woods or along fence rows to pick at least a few gallons in a day's blackberrying, making the planting of wild bushes unnecessary.

In my opinion, wild blackberries taste even better than cultivated ones. Following are several of the major species that can be found growing wild. To these can be added the many escapes from gardens that are also out there waiting for you. In short, there are both erect and trailing bushes, thorny and thornless kinds, black-red and amber-colored ones—all the myriad types that can be purchased from a nursery grow somewhere in the wild. Remember to watch out for thorns when gathering them—putting on both gloves and insect repellent isn't a bad idea:

Rubus flagellaris. A prickly or thorny trailing blackberry from which the dewberry was probably derived. Has white flowers and round or oblongish black fruits that ripen May–June. Native to eastern North America and hardy from zone 4 southward.

RUBUS HISPIDUS. The swamp blackberry or dewberry. A partly evergreen vinelike plant with no thorns but bristly perennial canes. It has sparse, small, white flowers and sour black fruits that ripen in May. Native to the east and hardy everywhere.

RUBUS MILLSPAUGHII. A thornless species that is not, however, the source of today's thornfree varieties. Many other blackberry species have thornless forms and it is from these that the thornless commercial types were bred.

RUBUS LACINIATUS. The cutleaf, evergreen, or Oregon evergreen blackberry. Native to Europe but naturalized on the Pacific coast this vigorous, productive species has thorny perennial canes that can either arch or trail on the ground. Its sweet, juicy, jet-black berries range from medium-sized to large and ripen June–July. Hardy from zone 3 southward.

RUBUS PROCERUS. Himalayan giant. Native to central Europe but introduced here in 1890 and has since become naturalized. A perennial stemmed creeping blackberry with very prickly stems, it produces large, black, thimble-shaped fruits in June–July. Hardy from zone 3 southward.

RUBUS ALLEGHENIENSIS. Mountain blackberry. A North American species growing 3 to 6 feet tall with slender berries twice as long as they are wide. Berries are often amber-colored and have a spicy flavor. Hardy from zone 4 southward.

THE BEST BLACKBERRY VARIETIES TO GROW

White blackberries . . . "tree" blackberries . . . thornless blackberries . . . ornamental blackberries . . . virus-free blackberries . . . blackberries that grow practically in the tropics and in subzero climates—probably no fruit has been so improved in the last decade as the blackberry. Blackberry varieties have been developed for practically every section of the United States except the extreme South, that area represented as zone 10 on climate maps—and even there blackberries bred for zone 9 often do well. As noted, there are two main types, *erect* and *trailing,* which differ in the character of their canes: *erect blackberries* having arched, self-supporting canes, and the *trailing types* having weak canes that must be tied to poles or trellises. These two types also differ in that trailing blackberries usually ripen earlier and are larger and sweeter than erect types, although they aren't as hardy. A few varieties are called *semi-trailing,* but they are essentially erect types because after the first year they generally become more or less arched in habit.

The following lists will indicate the best blackberry varieties for different areas, classifying the types of plants and describing fruit size, flavor, and color—for, as noted, there are black, red, yellow, and even white blackberries. Sources are only indicated here where the varieties are hard to come by (see Appendix III for general blackberry sources). Most of these bushes are very productive, often yielding 6,000 pounds of fruit per acre, and 6 plants are more than sufficient for a family of four if you don't plan to put up a large amount. The ripening times given—very early, early, midseason, etc.—show when a variety ripens in relation to other varieties grown in the same garden. According to the USDA, "the time lapse between ripening of very early varieties and very late varieties may be as little as 2 days or as much as 40 days."

Erect

(Generally easier to care for than trailing varieties)

ALFRED. An old variety developed in Michigan that does well there and in other northern states. Bushes are hardy, vigorous, productive, and early-fruiting, yielding large, firm, sweet blackberries. The plant's main drawback is that it is susceptible to orange rust. Hardy in zones 4, 5, and 6, though it should be protected in zone 4.

BAILEY. Another old variety, developed in New York and primarily grown there. Passing out of use today, this subacid, juicy, midseason berry is large, glossy black, medium-firm, and grows on tall, vigorous, productive plants that are also susceptible to orange rust. Hardy in zones 5 and 6.

BRAZOR. Developed in Texas in 1959 and usually grown there and in the Gulf Coast area. An early, very vigorous, productive plant that bears attractive, large, black berries that are fairly firm. Produces very well in a limited space. Hardy in zones 8 and 9, but can be grown farther north with protection just as boysenberries can. Available from Well's Nursery, Box 146, Lindale, Texas 75771.

DALLAS. This Texas-born variety is also much grown in Oklahoma. The plant is semi-trailing, vigorous, and productive, yielding early, large, firm, black berries. Hardy to 10° F. with winter protection, it does well in zones 5, 6, and 7.

DARROW. A commercial variety, developed in New York and much grown in the North, which is considered by many to be the best all-around blackberry. Plants are vigorous and very productive, bear early, and have a comparatively long fruiting season. The very flavorful berries are large, firm, glossy black, and mildly subacid. Hardy in zones 5, 6, and 7, and has survived temperatures of −20° F. Available from Boatman's Nursery, Bainbridge, Ohio 45612.

DESOTO. A vigorous grower that yields large, very sweet blackberries from midseason to fall. Good for northern areas.

EARLY HARVEST. Developed in Illinois and grown as far north as Maryland and southern Illinois, but very productive in the South and thus a desirable variety for that region. An early-bearing plant with a long fruiting season that doesn't develop many suckers and yields firm, medium-sized, black berries. Hardy in zones 6, 7, and 8, and particularly adapted to higher elevations there.

EARLY WONDER. A leading variety in its Texas home and Oklahoma. The semi-trailing, vigorous, productive plant bears firm, medium-sized, black berries and is hardy in zones 6, 7, and 8. Does well at higher elevations.

EBONY KING. A new, midseason, hardy type for the North and East that yields large, sweet, black berries. Available from Boatman's Nursery, Bainbridge, Ohio 45612.

ELDORADO. Also called Stuart, Lowden, and Texas. Developed in Ohio, it does well there but is a poor choice for the extreme South. This very vigorous, productive plant bears early to midseason, boasting a long fruiting season and yielding large, firm, sweet, black berries that are virtually coreless. The most resistant to orange rust of widely grown varieties. Hardy in zones 5, 6, and 7 and good for mountain regions there.

FLINT. Developed in Georgia and well adapted to the South and Southwest. Yields large, firm, good-flavored midseason black berries on productive, very vigorous, semi-upright plants. Resistant to leaf spot and anthracnose. Hardy in zones 7 and 8.

HAUPT. A late variety that originated in Texas and is often grown in the eastern part of that state. A productive semi-trailing plant that roots at the tips of the canes. Yields large, fairly firm, black berries and is hardy to 0° F. if protected from wind.

HEDRICK. Developed and presently grown in New York, this early variety bears large, glossy-black, juicy berries of medium firmness and irregular shape. Plants are tall, vigorous, and productive but susceptible to orange rust. Hardy in zones 5, 6, and 7 and well adapted to the mountain regions of zone 7.

HUMBLE. A midseason southern variety that is vigorous, productive, and yields medium-sized black berries.

JERSEYBLACK. Another midseason variety, developed in New Jersey, that has a long fruiting season. It is a vigorous, productive, semi-trailing plant bearing large, firm, mildly subacid black berries. Hardy in zones 6 and 7.

LAWTON. Also called New Rochelle. A midseason variety that boasts large, soft, sweet berries. The vigorous, productive plants are resistant to orange rust. Developed in New York but grown mostly in Texas and Oklahoma. Hardy in zones 6, 7, and 8 and does well in the mountainous regions of zone 8.

NANTICOKE. Also called Hirschi and Healthberry. Developed in Maryland and well adapted along the Atlantic coast, this is a very late variety with a long fruiting season. Its berries are of medium size, soft, and sweet, the plants vigorous, productive, and drought-resistant but very thorny.

Hardy in zones 6, 7, and 8.

RAVEN. An early berry with a short ripening season developed in Maryland but grown from southern Pennsylvania and southern New Jersey southward and west to the south-central United States. The moderately vigorous plants are productive in the South and yield medium-large and firm berries with a very good subacid flavor. Hardy in zones 6 and 7, but needs winter protection in zone 6.

SNYDER. A good cold-resistant variety for the North. Plants are vigorous and very productive, the berries medium-sized, very seedy, but with hardly any core.

Typical fruit cluster is (right) trailing blackberry variety and (left) erect blackberry variety.

Trailing

(Generally ripen earlier and bear larger and sweeter berries than erect types)

BRAINARD. An old standby developed by the USDA as a cross between the previously mentioned Himalaya species (*Rubus procerus*) and an eastern plant to provide the fine, large, juicy, black fruit of Himalaya on more sturdy canes. Late ripening and hardy in zones 6, 7, and 8.

CASCADE. Another old favorite, this one developed in Oregon and grown mostly in western Oregon and Washington. Plants are vigorous and very productive, early ripening, and yield large, bright, deep-red berries with excellent flavor. Of the highest dessert quality, Cascade most nearly approaches wild berries in flavor. Hardy only in the Pacific coast area (zones 7 and 8) and should be protected winters in zone 7.

CHEHALEM. A late variety developed in Oregon and adapted to the Pacific coast. Vigorous, very productive plants that yield small, shiny, black fruits with small seeds and excellent flavor. Hardy in zones 7 and 8 and needs winter protection in zone 7; not adapted to states east of Arizona.

FLORDAGRAND. Developed in Florida and best adapted there and along the Gulf Coast. The plants are vigorous, productive, and disease-resistant, especially to leaf spot and double blossom, but require cross-pollination. Berries are very early ripening, large, glossy, soft, and aromatic. Hardy in zone 9.

GEM. A Georgia development well adapted in the Gulf Coast region. An early-bearing plant resistant to anthracnose and leaf spot that yields large, round, firm, black berries with good flavor. Hardy in zone 8.

MARION. A late berry developed in Oregon and adapted to western Oregon and western Washington. Plants very vigorous and productive, with heavy canes. Berries medium to large, bright black, and of medium firmness. Hardy in zone 8.

MCDONALD. Widely planted in Texas, this old standby is self-sterile and another variety must be planted nearby it to ensure pollination (plant one Early Harvest for every five McDonald). Very early, good-quality fruit on a productive plant.

OKLAWAHA. Developed in Florida and suitable for that state and the Gulf Coast region. Very early black berries of medium size with good flavor and aroma. Serves as a pollinator for Flordagrand. Hardy in zone 8 and well adapted to high elevations there.

OLALLIE. An Oregon introduction very popular in California, western Oregon, and the Gulf Coast region. Very vigorous, productive plants, resistant to verticillium wilt, sunscald, and powdery mildew, but susceptible to rust and dry berry when grown in humid areas. Berries are large, firm, bright black, and sweet, tasting a lot like wild blackberries. Hardy in zone 8 and higher elevations there, and not adapted to states east of Arizona.

SANTIAM. Developed in Oregon and hardy only in the northwestern United States, especially Oregon's Willamette Valley. Bright, deep-red berries with excellent flavor on a plant of medium vigor that is susceptible to leaf spot.

Other good trailing varieties include Erie, hardy in zones 5, 6, and 7; Williams, hardy in zone 8; Mammoth, hardy in zones 7 and 8, but not adapted to states east of Arizona; and Dewblack, hardy in zones 6, 7, and 8.

Thornless

(Much easier to work with, but generally bear smaller crops)

BLACK SATIN. Introduced recently by the USDA and Southern Illinois University. A very early variety said to yield 20 to 30 sweet, juicy, black berries per stem. Plants semi-upright, highly disease-resistant, and does not sucker. Hardy in the Midwest and South and does well in the North with winter protection. Available from Lakeland Nurseries, Hanover, Pennsylvania 17331.

GEORGIA THORNLESS. This thornless Georgia development is well adapted to the Gulf Coast region. Productive, semi-upright plants that bear in midseason, yielding large, firm, oblong, black berries with a good flavor. Hardy in zone 9.

SMOOTHSTEM. Developed in Maryland by the USDA and well adapted for home gardens from southern Maryland to North Carolina along the Atlantic coast. Very vigorous, productive, nonsuckering plants with semi-upright, thornless canes that are resistant to leaf spot. Berries quite late, large, firm, tart, and good-flavored. Hardy in zones 7 and 8, and adapted to high elevations. Available from Raynor Bros., Inc., Salisbury, Maryland 21801.

THORNFREE. Another Maryland USDA introduction. Hardier than Smoothstem, taking temperatures of as low as −10° F., but still not winter-hardy in the North. Hardy from central New Jersey, southern Pennsylvania, southern Ohio, southward to North Carolina, and west to Arkansas, and in the Pacific Northwest. Plants medium vigorous, productive, thornless, semi-upright (more erect than Thornless), and resistant to leaf spot. Berries late, large, firm, and tart with a very good flavor. Available from Bountiful Ridge Nurseries, Princess Anne, Maryland 21853.

THORNLESS EVERGREEN. A mutation of the old variety Evergreen that is very popular in Oregon and Washington but not generally adapted to states east of the Rockies. Plants are vigorous, productive, very late in bearing, semi-trailing, deep-rooted, drought-resistant, and are propagated from tip plants only. Thornless Evergreen is susceptible to rosette in Atlantic coast states. Hardy in zone 8 and will stand temperatures of −10° F. if protected.

Other thornless varieties include Austin Thornless and Cory Thornless, both hardy in zones 7 and 8, but needing winter protection in zone 7.

Unusual Blackberries

WHITE BLACKBERRY. Creamy-white colored, mild-flavored, midseason berries. Available from Zilke Brothers, Baroda, Michigan 49101.

EVERBEARING TREE BLACKBERRY. Large, juicy, black berries over a long period on a strong-caned "tree" that grows 6 to 8 feet high. Available from Mich-O-Tenny Nurseries, Bainbridge, Ohio 45612 or Zilke above.

COLOMBIAN BLACKBERRY. A South American variety that bears large berries often 2 inches long and just as thick. Not available from U.S. nurseries.

VEITCHBERRY. A cross between a blackberry and a raspberry that bears very large, sweet fruits of a deep mulberry color and has the flavor of both parents. The semi-erect plant, which ripens its berries after raspberries have finished fruiting and before blackberries begin, should be cultivated like the blackberry.

KINGS ACE BERRY. Another cross between a raspberry and blackberry that produces large, sweet fruits. The semi-erect plants are not rampant and are very suitable for the home garden. This is one berry classified and treated as a blackberry, even though the fruit comes away easily from the receptacle like a raspberry.

WHEN, WHERE, AND HOW TO PLANT

While there are varieties for every region, blackberries do grow best in temperate climates, the plants usually not adapted to areas where summers are hot and dry or where winters are severe. After choosing a hardy, semi-hardy, or tender variety, according to the climate where you live, select a deep, moist, fairly rich loam in which to plant. If such soil isn't available, however, don't be afraid to plant in practically any garden soil; blackberries aren't fussy, except that the shallow-rooted plant doesn't like sandy soils and prefers a moist site that has good drainage.

Though the blackberry is cosmopolitan as to where it grows, an ideal blackberry soil can be recommended. This would be a loam with a pH of 5.5 to 7.5 that contains more clay than sand, one that while it is moist never becomes waterlogged and while it is fertile isn't so rich in humus that it supports vegetative growth at the expense of fruiting. As for an ideal site, a well-drained, cool, northern exposure is best. Try to protect the plants from drying winds, setting them in a sunny location sheltered by hills, trees, or shrubs. Where winters are severe, plant blackberries on hillsides; there they are in less danger of winter damage from late spring frosts than if planted in valleys. Additionally, to avoid possible diseases, try not to plant blackberries near wild species and to keep them far removed from any raspberries in the garden.

From zone 6 south, blackberries can be planted in the fall, winter, or spring; in all regions north of this, set the plants out in early spring as soon as the ground can be worked. Before planting, turn the soil over to a depth of about 9 inches. At this time you can incorporate compost, leaf mold from nearby woods, manure with a lot of straw in it, even grass clippings. Commercial fertilizers can also be used, but select one that isn't high in nitrogen (5-10-5) so that excessive leaf growth isn't encouraged. An alternative is to sow the ground to be planted with a green manure crop like rye or cowpeas the summer before and turn it under prior to planting. Of course, no fertilizing is necessary if the area for the blackberry plantation is fertile enough to begin with. But remember that blackberry plants are long-lived, a plantation producing fruit for up to twenty years; time spent in preparing the ground properly is time well spent.

Try to get blackberry plants into the ground as soon as you receive them from the nursery, taking care not to let the stock dry out. If planting isn't possible immediately, prevent drying out by heeling in the plants; that is, dig a trench deep enough for the roots and cover them with moist soil. Should the plants be dry when you receive them, soak the roots in water a few hours before planting. In any case, trim away all long roots and cut the canes back to about 6 inches. Then dip the roots in thin mud; this "puddling" will prevent the plants from drying out while you are setting them. Next, dig a generous hole, spread out the roots fanwise, and set a plant in the hole about as deep as it grew in the nursery, filling the soil back in, packing it down firmly with your foot around the collar of the plant, and then watering to further prevent air pockets from forming around the roots.

Erect or bush-type varieties of blackberries should be set 5 feet apart in rows 6 to 8 feet apart, except when planting hedgerows, in which the plants are set 2 feet apart in rows 8 to 10 feet apart. The more vigorous-growing trailing and semi-trailing types should be spaced 8 to 12 feet apart in rows 10 feet apart. Don't skimp on space, or picking the thorny varieties will be much harder. During the first year intercrops like cabbage, beans, peas, and summer squash can be grown between blackberry rows, but in subsequent years the plants will need all the nutrients and moisture in the plantation and no intercropping should be practiced.

Pruning And Thinning

The pruning of blackberries is best understood when the growth habits of the plant are known. Blackberry plants are perennial, but new canes or stems arise from them every year. These biennial canes—they live only two years—send out lateral or side branches and in the second year smaller branches grow from the buds on the laterals. Fruit is borne on these branches and after the laterals bear fruit, the canes die.

Accordingly, the first object in pruning blackberries is to develop lateral branches. *The first step* is to cut off the tips of new blackberry canes of the current year's growth when they reach a height of 3 feet (8 feet for trailing varieties). This tipping will both make the canes branch and cause them to grow stouter and better able to support a heavy crop of berries. *The second step* is to cut the laterals or side branches that develop on these stout canes back to about 12 to 24 inches *before growth starts early the next spring*. These two practices are essential to ensure heavy fruiting and better-quality fruit.

It is also very important to thin out new blackberry canes when they begin to grow every spring. Weak canes should be removed and others thinned so that there are no more than 12 to 16 canes on each plant for trailing varieties, 4 to 8 for semi-trailing varieties, and 3 to 4 for erect varieties. The excess canes are removed by what is called "rubbing," that is, cutting them off at ground level.

In addition to new canes, suckers are also sent up by blackberries every year. If all of these were allowed to grow, the blackberry planting would become almost impenetrable. During the season, remove all of the suckers that appear between the rows by pulling them out of the ground—suckers that

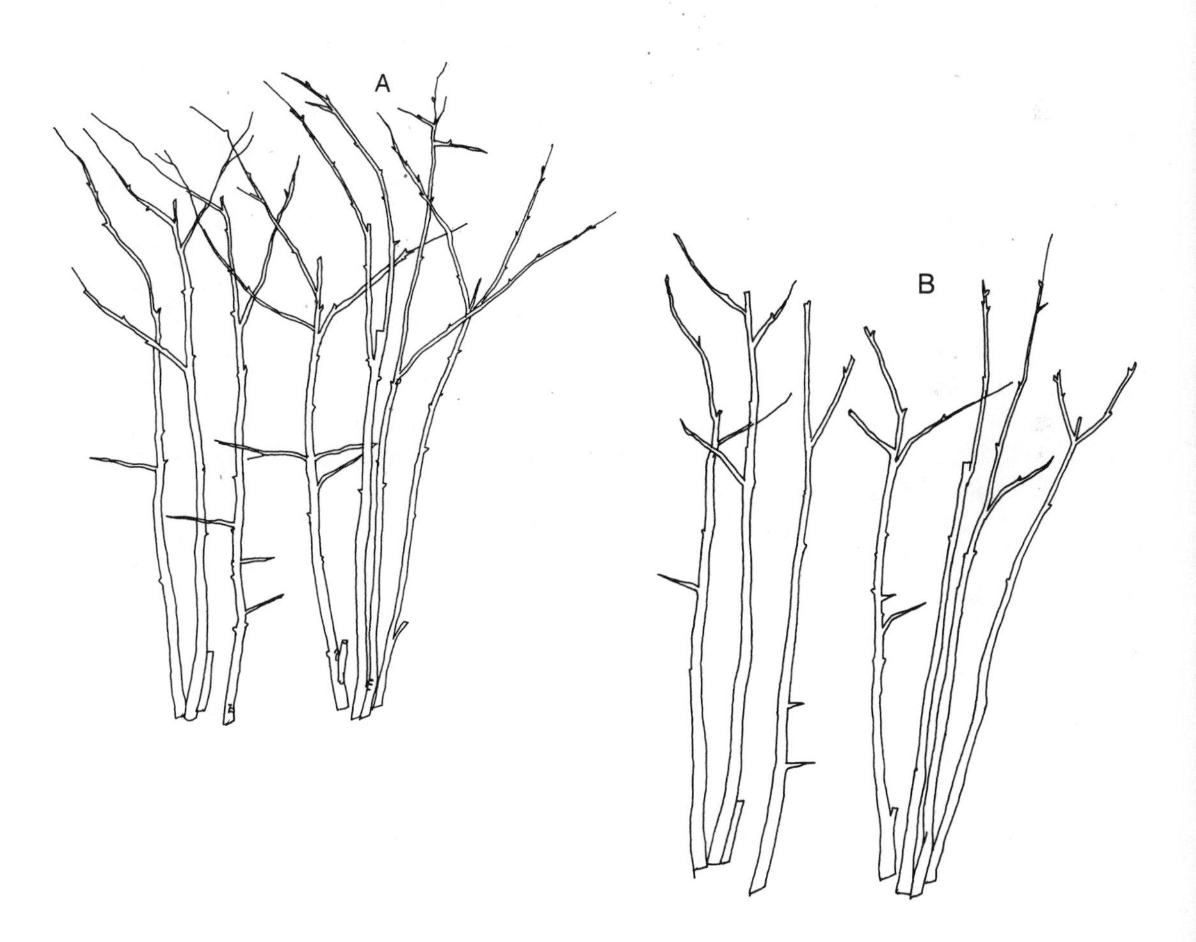

A blackberry plant before (A) and after (B) pruning.

are pulled will not regrow as fast as those that are cut down. If you want to let some suckers grow *within* the row, thin them to about five or six canes per lineal foot of row.

A final pruning task is to remove all injured canes and destroy all canes that have fruited at the end of the season. This is a sanitary measure that helps prevent the spread of disease in the planting, makes the garden look better, and permits fuller development of young canes.

When doing any pruning in the blackberry patch it's a good idea to wear heavy gloves and a long-sleeved shirt and to use long-handled pruning shears.

Space-saving Ways to Train Blackberries

Many gardeners grow *erect* blackberry plants without any support at all and this is a perfectly good method, but since the canes can be broken during cultivation and picking, it is a good idea to train even erect varieties on wires, against a fence, or over a pergola. As for trailing-type blackberries, these can be allowed to sprawl if spread properly and are sometimes used as a ground cover for difficult banks. They are, however, much easier to work with if trained and produce bigger berries if treated this way. There are many ways to train blackberries, saving space and your sacroiliac at the same time, and you may want to experiment with new ideas after growing them a few years. But here are the basic methods for both erect and trailing varieties:

• *Erect blackberries* are tied to wire attached to two sturdy 5-foot posts set 15 to 20 feet apart in the row. Stretch a single galvanized iron wire taut between the posts about 2½ feet from the ground. Then just tie each cane with soft string at the place where it crosses the wire. Do not bend the canes —tie them as they stand upright—and don't bunch them together and tie them.

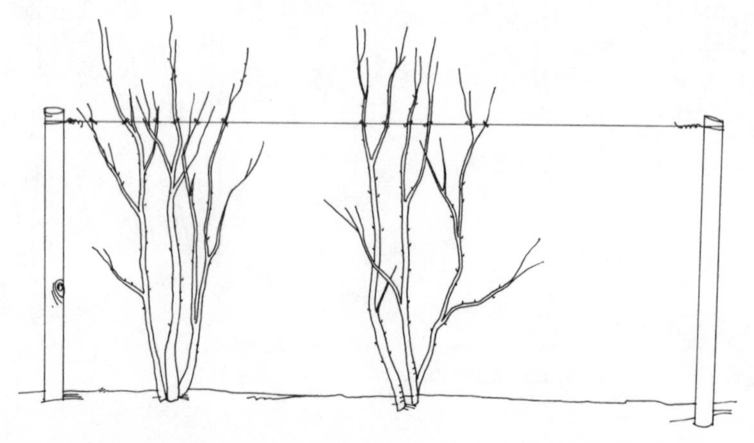

Train erect blueberry varieties to a one-wire trellis.

• *Trailing blackberries* are also trained on two sturdy 5-foot posts, the posts set at the same distance apart in the row. Here, however, *two* wires are attached to the posts—one about 3 feet from the ground and the other about 5 feet from the ground. Tie each trailing cane horizontally along the wire or fan each one out and tie it where it crosses each wire. Again, don't tie the canes in bundles. Young shoots of the current season's growth (canes that will fruit next year) can be kept out of the way by stretching them out along the ground at the foot of the trellis and holding them in place with brackets or forked branches.

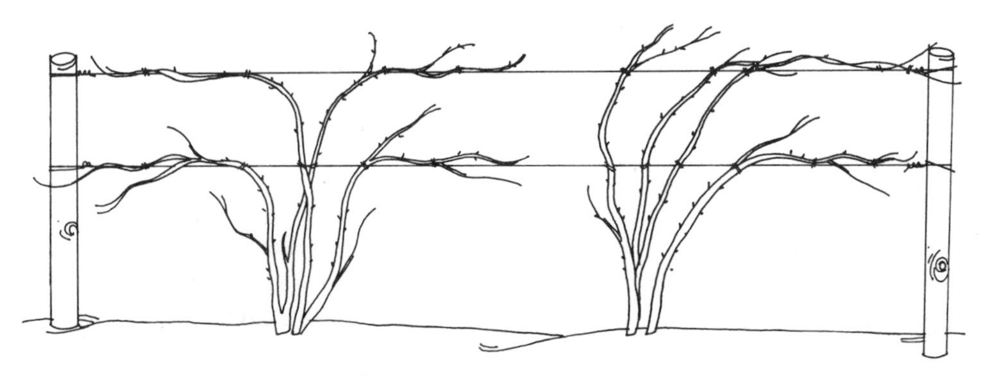

Trailing blackberries should be tied to a two-wire trellis.

Care Of Blackberries

Fertilizing and Watering

Commercial growers, whose plants yield up to five gallons of blackberries each, usually fertilize twice a year. At blossoming time they apply 5-10-5 fertilizer as a top dressing at the rate of 500 to 1,000 pounds per acre or 5 to 10 pounds per 50 feet of row. After fruiting they apply nitrate of soda at the rate of 200 to 300 pounds per acre, or ammonium nitrate at the rate of 80 to 100 pounds per acre.

Some growers also apply a dressing of chelated iron where there is lime in the soil and the plants suffer from iron deficiency (indicated by yellowing leaves).

Organic fertilizers often used for blackberry plants are fish and bone meal. But more important is sowing a cover crop or using a mulch. Humus and moisture in the soil are by far the most important factors in blackberry fruit production and both of these methods provide them. Mulches, explained in the entry following, are more appropriate for the home gardener with a smaller planting. Cover crops, for the big grower, should be sown after the picking season is over. Barley, oats, buckwheat, and millet are sown by some gardeners and turned over in the fall, which also helps to check succulent growths that would be winter-killed. Others sow cowpeas, spring oats, or rye between rows at the last cultivation, leaving 12 to 15 inches next to the plants free of cover crop, and turn under the cover crop the following spring. Besides adding humus to the soil, this helps prevent soil erosion and protects canes during the winter by shielding them from drying winds.

Remember to keep plants watered throughout the growing season if a mulch isn't used—especially with first-year plants and during the 3 weeks before ripening.

Keeping Weeds Down: Cultivating, Mulching, and Herbicides

Weeds must be kept down in the blackberry patch, but great care should be taken in cultivating blackberries because of their shallow roots. Beginning in the spring, as soon as the soil is workable, cultivate only 2 to 3 inches deep near the rows. Keep hoeing every 1 to 2 weeks all summer until about a month before the first frost is expected. If you continue cultivating beyond this time, new growth will be encouraged when the plants should be hardening off and they will be more prone to winterkill. If machine cultivators are used, a tractor-mounted grape hoe or rotary is better than a shovel cultivator.

Many gardeners cultivate only until midsummer and then mulch their plants with 4 to 6 inches of old hay or 2 to 3 inches of sawdust, wood chips, ground bark, grass clippings, or whatever material is available (see Appendix II for an extended list of good mulches). Still others eliminate cultivating altogether and mulch the plants all year around to keep down weeds. Mulching not only conserves the moisture so essential to blackberries (moisture that weeds rob), but has the added benefit of feeding the plants as well.

Herbicides are another weapon in the war on weeds, but can be more dangerous than the weeds themselves, especially if used imperfectly. I have never found them necessary, encountering no problems with a good mulch, but I list them below in case you decide to use them:

• *Simazine.* Will not control established weeds, but is applied to the soil to control pigweed, lamb's-quarters, crabgrass, and many other broadleaf weeds and grasses. Use after clean cultivation of the soil in early spring.

• *Diuron.* Same control as Simazine. Use after clean cultivation during the dormant period, after canes have been trained. Growers often rotate

Simazine and Diuron from year to year to avoid excessive buildup of herbicide residue in the soil and prolonged exposure of the crop to one herbicide.

• *Dinoseb.* Used during a dormancy to kill young weeds less than 4 inches tall.

• *Chlororopham.* A dormant treatment that controls both germinating and established chickweed.

• *Dichlobenil.* Applied during cool weather in late fall or early spring (and only on established plantings) to control annual and certain perennial weeds. If applied when new shoots are emerging this herbicide will cause temporary stunting of the new shoots.

Be sure to read directions on all herbicides carefully.

Insect Pests and How to Beat Them

Blackberries are strong, vigorous plants that aren't as likely to suffer from pests as raspberries or others of the bramble fruits. But they do occasionally fall victim to such troubles. Blackberry fruit or foliage may become infested with insects like aphids, beetles, fruitworms, leafhoppers, leafrollers, mites, sawflies, stinkbugs, and weevils. Usually, these can be controlled with three applications of one of the many multipurpose fruit sprays on the market: one application when the leaves unfold in spring; another when new canes grow 6 to 8 inches high; and a third just before flower buds open. Other general precautions include pruning out insect-infected canes and burying them, removing old canes after harvest, and keeping the garden clean. But there are specific controls for specific pests if any one proves particularly troublesome, as the following list shows. Chemical controls are noted first—*and the manufacturer's instructions should be followed carefully when applying them*—but it would be a good idea to try the organic controls suggested before using these chemical preparations.

APHIDS. Soft, tiny, green lice that cluster on leaves and curl and distort them. Dust or spray with Diazinon or malathion or parathion. There are also several organic remedies such as rinsing plants with soapy water, planting aphid-repelling rhubarb, nasturtiums, garlic, or chives nearby, and controlling the ants that carry aphids by sprinkling bone meal near their holes.

JAPANESE BEETLE. A half-inch-long, shiny, green and bronze insect with brown wings that feeds on foliage. Spray with malathion or carbaryl or methoxychlor. Or handpick; spray with organic rotenone; plant odorless marigolds, larkspur, and white geraniums nearby to attract beetles away from raspberries and collect the insects from these trap plants; use the bacterial insecticide Doom, which transmits the fatal milky spore disease to Japanese beetle grubs (larvae) and breaks their breeding cycle.

ORGANIC TORTRIX. A small, orange-colored insect that damages leaves. Spray with carbaryl.

RASPBERRY CANE BORER. A small, black, long-horned beetle with a yellow neck that girdles tips of canes and whose larvae bore down in pith of canes, causing them to wilt. Spray with organic rotenone; cut off infested canes and burn; plant garlic nearby to repel the borers.

RASPBERRY CROWN BORER. An insect resembling a yellow jacket that enters canes (especially on trailing varieties) leaving no sign of entry and hollows them out, causing cane growth to be weak. Spray with Diazinon or parathion. Destroy seedlike borer eggs laid on undersides of leaves; try organic rotenone spray.

RASPBERRY FRUITWORM. Small, light-brown beetles that feed on raspberry flowers, their slender grub offspring boring into the fruit. Spray with organic rotenone.

RASPBERRY SAWFLY. Small, pale-green, many-legged worms that feed on leaf margins. Spray with organic rotenone.

RASPBERRY MITE. Very small mites that damage raspberry leaves. Sprinkle on lime and sulfur; direct a jet of plain water at mites; try organic rotenone spray.

RED-NECKED CANE BORER. Shiny blue-black beetles with a red neck or thorax that eat leaf margins, their slender, whitish larvae or grubs boring upward in canes and causing cigar-shaped swellings or galls. Spray with organic rotenone; cut off and destroy infected canes well below the point of infection.

RED SPIDER MITES. See Spider Mites, below.

ROSE CHAFER. Medium-sized, yellowish, long-legged, hairy, clumsy beetle that eats foliage and flowers while its larvae or grubs attack roots. Spray with carbaryl or methoxychlor; sprinkle with rose dust every week.

ROSE LEAFHOPPER. Small, yellowish-green to brown, often spotted, wedge-shaped insects that cluster on undersides of leaves (flying when disturbed) and suck juice from the leaves, transmitting yellows to the plant. Spray with malathion; dust with sulfur or spray with organic pyrethrum or rotenone.

ROSE SCALE. Various sizes of round or elongated white, brownish, or purplish waxy scale under which tiny insects live clustered on the leaves and stems of plants. Spray with malathion; use a lime or sulfur spray.

SNOWY TREE CRICKET (or TREE CRICKET). Medium-sized, pale-green, slender cricket that punctures canes to deposit eggs, breaking them, and feeds on leaves, flowers, and fruit. Remove affected stem and burn.

SPIDER MITES. Minute, reddish-green forms that suck juice from leaves, causing them to yellow, and make their webs over leaves. Spray with Demeton or Dicofol. Or hose insects off plants with strong jets of water; use onion spray (onions crushed and mixed with water); dust with organic pyrethrum.

STINKBUG. Medium-sized, green, black, or brown, shield-shaped bug with an unpleasant smell that sucks juice from foliage and flowers. Spray with carbaryl. Or keep plants clean of weeds in which stinkbugs gather and handpick the bugs from plants (wearing gloves).

STRAWBERRY WEEVIL. Small, almost black beetle with short snout that feeds on leaves and berries and whose white, legless, curved larvae or grubs feed on crowns. Spray with methoxychlor.

TARNISHED PLANT BUG. Medium-sized, dirty-green, or tan bug that flies when disturbed (the nymphs are wingless) and causes imperfect berries. Spray with Sevin when flowers first open.

TREE CRICKET. See Snowy Tree Cricket, above.

WHITE GRUBS. Adults are big, clumsy, reddish-brown, oval beetles which have large larvae or grubs that are yellowish-white, black-headed, and six-legged and curl up in the soil, attacking roots. Apply chlordane.

Diseases and Remedies

The blackberry also shares most of the same diseases as the rest of the bramble fruits and raspberries. Generally speaking, these afflictions can be kept at a minimum if the following precautions are taken:

- Immediately destroy all disease-ridden plants observed.
- Remove all wild blackberry plants in the nearby area.
- Keep the planting free of weeds and fallen leaves.
- Remove old canes from the planting every year.
- Plant other crops for several years before replanting an area to blackberries.
- Select disease-resistant varieties for your region.
- Choose high-quality planting stock free of nematodes and diseases.

Regarding specific blackberry diseases, their symptoms, and their remedies, the following list of the 12 worst diseases will be of help:

ANTHRACNOSE. Caused by a fungus that produces purplish to gray sunken spots on canes, leaves, and fruit stems; most severe in Southeast. To control: practice good sanitation; enrich soil and improve drainage; protect against rain-splash by mulching properly; remove all suspicious-looking plants; dust with Zineb; spray with lime sulfur spray in late winter when plants are dormant.

CROWN GALL AND CANE GALL. Here soil bacteria infects plants through wounds made by pruning and cultivation, producing warty galls on crowns and canes and weakening plants. Cultivate and prune carefully; plant gall-free nursery stock.

DOUBLE BLOSSOM. A fungus disease that makes flower buds enlarge and produces coarse, reddish flowers with twisted petals (the flowers appearing to be double); flowers will not produce fruit when struck by this disease most common to the Southeast, and infected plants also develop abnormal, broomlike growths of leafy shoots. If the disease strikes, cut *all* canes close to the ground after harvest and remove them.

FRUIT ROT. Caused by several fungi that infect the fruit before or after harvest, coating it with a gray or black cottony fungus mass; most prevalent on damaged and overripe berries. As a control apply captan at the beginning of bloom and at 10-day intervals until harvest (if fruit rot is severe, as in excessively rainy weather). Carefully pick only sound fruit at frequent intervals and immediately store it under refrigeration at 32° F. to 40° F.

LEAF CURL. A disease spread by aphids that dwarfs plants and curls their leaves, making them unproductive. Use controls for Aphids (see under Blackberry Insect Pests, in the Index).

LEAF AND CANE SPOT. A fungus disease producing irregular purple and light to dark-brown spots on leaves and canes (older spots are whitish with reddish margins) that causes early defoliation and weakens plants. Most severe in humid areas in Southwest and Northwest. Cut out infected canes immediately after harvest; spray same as for Anthracnose; in Pacific Northwest spray with Bordeaux mixture (8-8-100).

MILDEW. This fungus disease coats leaves with a gray powdery substance, turning leaves upward, and weakening plants. Apply lime sulfur spray as leaf buds begin to swell, and repeat when new canes are 6 to 8 inches high.

MOSAIC. Another disease spread by aphids, in which leaves become mottled light and dark green, leaves curl downward, and plants become dwarfed and unproductive. Same control as for Leaf Curl, above.

ORANGE RUST. A fungus disease established in the crown of the plant that produces orange or yellowish masses or pustules (spores of the disease) on both surfaces of leaves, making leaves narrow and plants stunted and barren of fruit. No effective control except to buy resistant varieties and plant disease-free stock. Immediately dig out infected plants, roots and all, before the spores are discharged to infect other plants. Destroy nearby wild plants.

SPUR BLIGHT. Reddish lesions that occur around cane spurs causing buds to die or be severely retarded; infested areas turn gray during winter. Spray with Bordeaux mixture 7 to 10 days before bloom and repeat as blossoms start to open.

STERILITY (no fruit). Can be caused by several factors, including the tarnished plant bug, which feeds on blackberry flowers causing them to develop into sterile or partially sterile fruits (see Blackberry Insect Pests, in the Index, for controls), but is usually the result of a virus disease that causes plants to produce completely or partially sterile flowers. The disease is common in the East and infected plants grow without apparent symptoms until fruiting time, when only a few druplets develop on each fruit. Other symptoms are the production of new canes that are more vigorous with rounder and glossier leaflets and premature reddening of leaves in the fall. This virus spreads to nearby plants. The only controls are to root out infected specimens and destroy them; remove canes from all plants after harvesting; destroy wild blackberries in the area; stop using root suckers to propagate plants where the virus is present (suckers spread the disease); and, above all, buy only guaranteed, virus-free stock from nurserymen.

VERTICILLIUM WILT. A common soil-borne fungus that infects canes of susceptible varieties (such as Lucretia, the boysenberry, and the youngberry), causing them to turn yellow and die in midsummer. Controls are to plant in ground free of the fungus; use resistant varieties (Evergreen, Lawton, Olallie); keep the garden clean; and don't grow blackberries in fields that have recently grown potatoes, tomatoes, peppers, eggplant, and other solanaceous crops. Chemical fungicides do little good here because the verticillium fungus lives so long in the soil, but costly fumigation with chloropicrin does control the disease.

Always follow the manufacturer's instructions carefully when using any of the chemical sprays mentioned above.

Winter Protection

Many blackberry varieties are winter-hardy in the regions where they are grown. But a number of varieties I have indicated here need winter protection in certain zones, and northern gardeners can grow varieties suitable for warmer areas if the plants are protected during the winter. A regular mulch helps protect plants in cold climates, but in areas where winter temperatures are colder than −20° F. even the hardiest varieties should be bent over to the ground and completely covered with a layer of straw, cornstalks, coarse manure, mulch, or soil, and not uncovered until spring, when severe weather has ended. The same applies to varieties that are grown farther north than their usual range—and here only experimenting will tell if the plants survive. Erect plants that are trained on wires should be taken down when they must be winter-protected. Bushlike plants should have their side branches pruned to about 12 inches at the end of the season to make prone winter protection easier.

Cold weather isn't the only problem blackberries encounter in the winter; they suffer from drought as well. Because persistent winds dry the canes out in the Plains states the same protection must be given there and the soil must be well soaked before winter approaches as well.

In areas like western Oregon, where winters are mild and moist, the canes of trailing varieties sometimes suffer winter injury if they are allowed to lie on the ground. There the exact opposite form of winter protection is given—the trailing canes are tied to trellises at the end of the fruiting season and kept up through the winter.

Propagation

Blackberries are so easy to propagate that once you have a few plants you never need to buy more.

Erect or bush-type blackberries can be propagated from root suckers or root cuttings. *Suckers* are fast-growing stems that arise from the roots; they are dug up and replanted elsewhere. Use suckers about 6 inches tall and never take suckers from plantings where the blackberry sterility virus is present. *Root cuttings* can be dug from around established erect plants in early spring. The roots selected should be at least ¼ inch in diameter, about the size of a lead pencil. Cut them into 3 to 4-inch lengths and plant them in 2 to 3 inches of good soil about two feet apart in the row. Another method is to take the root cuttings the same way in the late fall, just before the ground freezes. Pack them in moist sand in a cool place where they won't freeze and keep them moist through the winter. Then in early spring, set these stratified root cuttings out in the garden. Root cuttings provide stronger root systems and many more plants than suckers, which aren't abundant in small gardens.

Trailing- and semi-trailing-type blackberries, including dewberries and loganberries, are best propagated by *tip layering.* In August insert the tip of a cane straight down into the loosened soil 2 to 3 inches. Hold the tip in place with a fork-shaped branch if necessary. Roots and latent stem buds will develop under the soil and the tip should be rooted within two months. If it is, it can be severed from the mother plant and transplanted to its permanent place. If not, wait until spring to sever and transplant the new plant. It should be noted that certain thornless plants, like Thornless Evergreen, must be propagated by tipping, as the other methods produce thorny plants.

Harvesting Blackberries

English legend has it that on October 11, Satan was expelled from Heaven and landed on a thorny blackberry bush. He damned the berries and from that day on they were good to eat only after October 11. In any event, blackberries will begin to yield the year after planting (on canes that grow the first year). The berries are not ripe when they first turn black. They must be harvested at just the right time—when they are sweet and ready to drop off at the slightest touch—for unripe berries have a sour taste and berries left too long on the canes taste spoiled or bitter and are too soft to keep well. The canes should be picked every other day when the berries begin to ripen and should be harvested early in the day when they do not spoil as easily. Be careful not to crush or bruise the fruit and keep the berries cool and dry after picking, never leaving them exposed to the sun. Wear the same thick gloves and protective clothing when harvesting blackberries that you wear when pruning them.

Blackberry Feasts

Use fresh blackberries right away, and never keep them longer than overnight in the refrigerator. Probably the best way to eat the berries is to serve them unchilled, sprinkled with sugar, and half covered with cream, but their deep, winy flavor also makes them wonderful for cooking, as many of the following recipes will prove.

Blackberries en Chemise

2 egg yolks
1 egg
½ teaspoon vanilla
3 tablespoons cornstarch
⅓ cup plus 1 teaspoon sugar
¼ cup milk
2 tablespoons kirsch
1 cup heavy cream
3 cups blackberries
¼ cup blanched almonds, toasted and chopped
2 tablespoons confectioners' sugar

Preheat broiler. Butter bottom and sides of shallow 6-cup heatproof dish. Put egg yolks, egg, vanilla, cornstarch, and ⅓ cup sugar in heavy saucepan over low heat and whisk together for a minute until thick. In separate pan, scald ¼ cup milk and gradually whisk it into yolk mixture, cooking and stirring it for a minute or two until it is thick. Remove from heat and gradually whisk in kirsch and heavy cream. Spread half of this thinned pastry cream in bottom of buttered dish, distribute blackberries evenly on top, and cover with rest of cream. Mix chopped almonds with teaspoon of sugar and sprinkle on top of cream. Distribute confectioners' sugar evenly over all. Slide under broiler, about three inches from heat, for three to five minutes, watching very carefully, until blackberries have just started to release their juices. Serve hot with chilled cream.

Blackberry Cobbler

Use the same recipe as for Strawberry Shortcake (see Index), substituting an equal quantity of blackberries for the strawberries.

Frozen Blackberries

To freeze blackberries to eat fresh later in the year, choose only plump, glossy-skinned fruit. They can be frozen unsweetened or in a sugar syrup made by dissolving three cups of sugar in four cups of water. Wash the berries in ice water and dry them thoroughly on paper towels before covering them with the cold syrup and packaging, sealing, and freezing them.

Easy Blackberry Wine

For each gallon of wine, crush two pounds of blackberries in a stone or glass crock (metal won't do) and add to one quart of water with two pounds of sugar. Boil the water first, adding the berries when it cools to lukewarm and covering the mixture loosely. Let stand for about a week or until most of the fermentation is finished. Then strain through cheesecloth into large jugs. Each jug neck should be covered with a toy balloon and let alone until all fermentation stops (in about a year), when the wine is tasted and bottled.

Simple Blackberry Jam

For each pound of firm blackberries used, add ¾ pound of sugar (1¾ cups). Put in a large pot over medium heat, stir, and remove from heat as soon as the first bubbles from the boil appear. Transfer the jam to glasses when it is cool.

Blackberry Cordial

"Soothing and efficacious in the summer complaints of children," says the old recipe for this. Directions are: "Take two quarts blackberry juice, one pound loaf sugar, four grated nutmegs, one quarter ounce ground cloves, one quarter ounce ground allspice, one quarter ounce ground cinnamon. Simmer all together for thirty minutes, closely covered to prevent evaporation. Strain through a cloth when cold and add a pint of the best French brandy."

Blackberry Pudding

Use same recipe as for Huckleberry Pudding (see Index), substituting an equal amount of blackberries for the huckleberries.

Pickled Blackberries

Combine one pound sugar, one pint vinegar, one teaspoon powdered cinnamon, one teaspoon allspice, one teaspoon cloves, and one teaspoon nutmeg. Boil all together gently, fifteen minutes, then add four quarts blackberries, and scald ten minutes more.

Blackberry Syrup

To 2 cups of blackberry juice add ¼ cup of white sugar, ½ ounce of powdered cinnamon, ¼ ounce of mace, and 2 teaspoons of cloves. Boil all together for ¼ of an hour, then strain the syrup and add to each 2 cups 1 shot glass of French brandy. Excellent for pancakes.

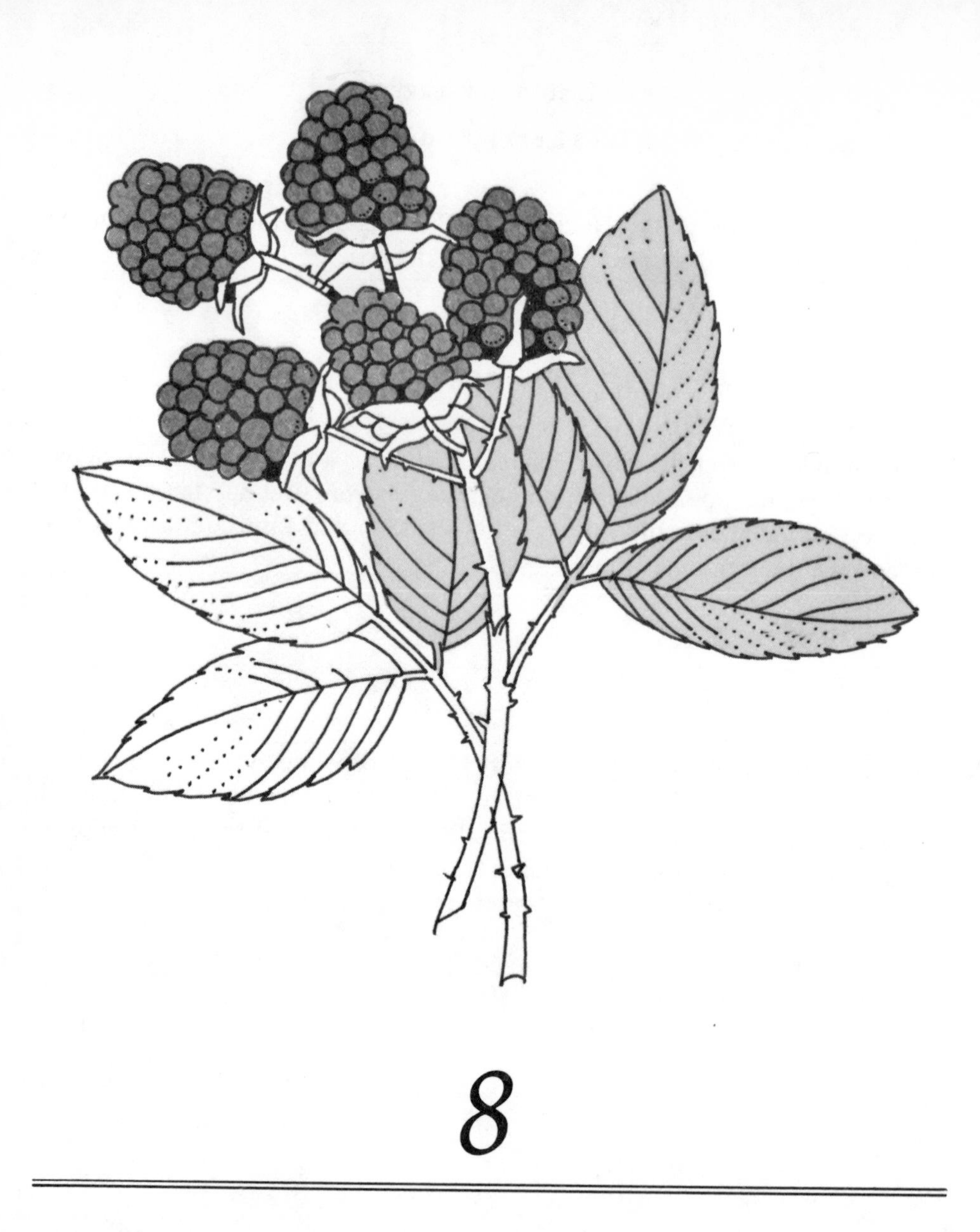

8

Boysenberry Bonanzas—
North and South

Americans have always been piemakers without peer, thanks to sugar resources close by, an abundance of native fruit, and a willingness to experiment. The blackberry, long regarded as a nuisance and called a bramble or brambleberry in England, is a case in point. As we've seen, many types of blackberries were developed here, long before anyone paid any attention to the genus *Rubus* in Europe. Among them is the boysenberry, a prolific prostrate or trailing variety that is a cross between the blackberry, the prostrate dewberry, and the raspberry (although some say it is a hybrid of the blackberry, the raspberry, and the prostrate loganberry).

No matter what its exact origins, the boysenberry is a relatively new fruit, having been developed by Anaheim, California, botanist Rudolf Boysen in only 1923. A dark, wine-colored fruit that has very few seeds and is extremely large for a berry—growing up to 2½ inches long and 1½ inches wide—it is bigger, richer, juicier, and tastes more "raspberryish" than the raspberry. Since Boysen "invented" the luscious "gourmet's berry" and it was first cultivated by Walter Knott of the famous Knott's Berry Farm in Buena Park, California, other hybridizers have improved upon it and today

Boysenberry fruit.

two major varieties are available: Thornless and Sensational, which is sometimes called The Wonder Berry. Thornless (offered by Field's, Burgess, May, Spring Hill, Hastings, Shumway, Boatman's, and Gurney's) is the most popular because it is completely free of thorns and easy to pick without scratching one's hands. Sensational (which can be ordered from Field's) does have a few thorns, but is more prolific and produces larger berries than Thornless. Both these varieties are vastly superior to older types of boysenberries, yielding up to ten quarts a plant for pies, jellies, juices, and eating out of hand.

Thornless boysenberry.

When, Where, And How To Plant

Boysenberries are generally regarded as a warm-climate fruit suited especially to the humid South, but they are hardy in zones 10 through 5 (any place where temperatures don't go below 5° F.) and can be grown north of Washington, D.C.—as far north as upstate New York, in fact—if protected properly through the winter. They will not bear as well in northern areas as they do in the South, but will yield a good crop if properly cared for.

Plant boysenberries in a sunny spot that isn't shaded by trees. They prefer a moist but well-drained soil that is slightly acid or neutral (ranging from pH5.5 to 7.0) and enriched with manure or compost. But if no other site is available they can be grown in fairly dry conditions and poor soil and still yield large crops. Since the trailing canes (they might better be called vines) grow as long as 8 to 15 feet, make sure there is plenty of space between plants for proper culture, setting them in rows at least 8 feet apart in the row. Plants should be allowed to grow sprawling the first year, for they won't fruit until a year after planting—on one-year-old canes.

Space Savers

If you choose to trellis the sprawling prostrate canes of boysenberries or train them against a fence or on wire supports, the plants can be set closer together. Using the latter method, wooden or steel posts are set 6 feet above the ground along the boysenberry row in early spring just before planting. Heavy steel wires are then stretched taut between the posts about 1½ feet apart, with the top wire about 6 feet high. The plants should be set about 8 feet apart, 1 foot away from the wires, but the rows can be 4 or 5 feet apart. The year after planting cut six or so canes back to about 6 feet long and tie them to the wires in a fan shape, cutting out any extra canes on a plant. After harvesting these fruiting canes at the end of summer, prune them out, for they will fruit no more, but don't tie up the new growth made during the summer, leaving these canes to sprawl on the ground until the following spring when they should be tied to the wires. The new canes can

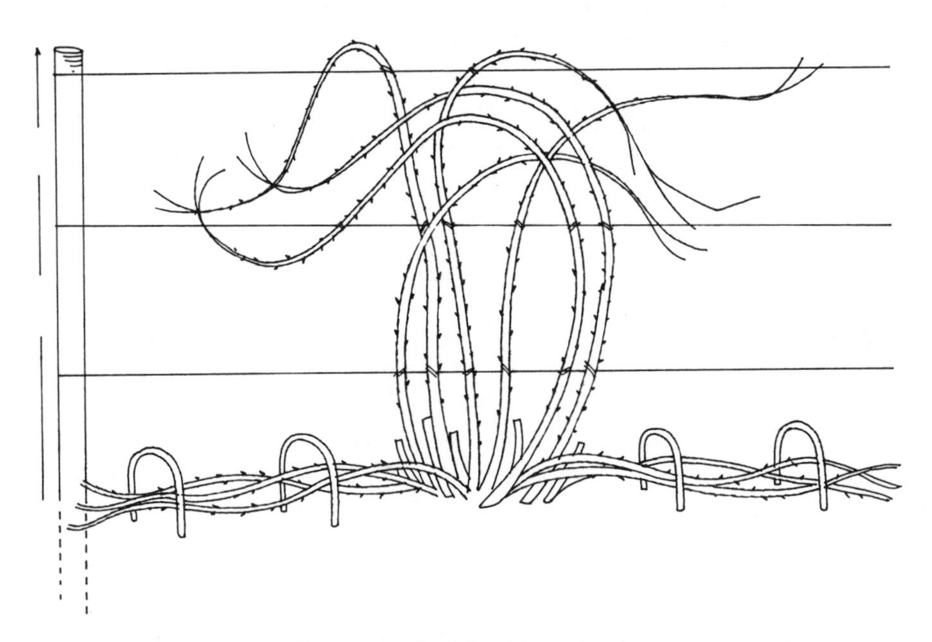

Training method for boysenberries.

be contained on the ground by holding them together in a straight line under brackets.

Boysenberries can also be tied up to 6-foot posts (one post to a plant) the year after planting, their ends cut off at the top of each post and fruiting canes being removed at the end of the season. No matter which training method is used, the berries will be bigger than if plants are allowed to trail along the ground. They will be easier to pick and cleaner as well.

CARE OF BOYSENBERRIES

Fertilizing, Watering, and Weeding

Fertilize boysenberries as soon as the buds begin to swell in early spring, taking care not to work the fertilizer in too close to the roots. Nitrate of soda or sulfate of ammonia at the rate of 1 pound per 100 square feet are good choices. So is ammonium nitrate, 2 ounces per plant, or a cupful of 5-10-5 fertilizer per 15 feet of row. Organic fertilizer for boysenberries would include 5 pounds of dried chicken manure or well-rotted compost for each 100 square feet.

A good mulch is another source of fertilizer for boysenberries, ensuring an adequate supply of water and holding down weeds in the bargain. Three or four inches of sawdust, wood shavings, or corncobs, or ten inches of straw, make fine fertilizer-mulches when mixed with cottonseed meal, dried blood, or any nitrogenous fertilizer. if nitrogen isn't added to these mulches, the yield of the plants will decrease. The mulch should not be applied until blossoms form on the plants, clean cultivation around the boysenberries being practiced until that time. Mulches will eliminate the need for weeding and only on the largest plantings will the use of commercial weed killers be necessary.

Winter Protection

Wintering over boysenberries in harsh climates can be done in several ways. Remember that it is next year's fruiting canes that you are protecting —not the roots, which cold winters rarely kill off. Sometimes snow alone is sufficient cover for canes in northern areas, but to be certain most experienced gardeners cover the sprawling canes with two or three inches of materials like straw or hay, or a foot or so of leaves. Canes that are tied on trellises must, of course, be unloosed and allowed to sprawl on the ground. All plants should be covered just before cold weather is expected to set in.

The trouble with protecting boysenberries in the above manner is that it doesn't provide sufficient warmth in really cold climates and frequently encourages the plants to tip-root under the covering, resulting in a jungle of canes for the gardener to contend with. Some growers therefore bunch together all the canes on each plant, cutting off an inch or so at the tips. Each plant or bunch of canes is then enclosed in a layer of bushy asparagus tops which is, in turn, wrapped up in a thick layer of dried cornstalks, and everything is tied up with strong twine in two or three places. The canes survive the coldest subzero winters in this fashion. There is no reason why other insulating materials could not be substituted for the asparagus tops and cornstalks. All coverings should be removed after hard freezes are no longer expected in the spring.

Pruning

Because boysenberry canes grow one year, bear fruit the second, and then die, canes must be cut out and disposed of in a sanitary manner just after fruiting. Aside from this, boysenberry pruning is just a matter of cutting out broken or injured canes, not letting the canes get tangled and out of control, and making sure that side shoots don't grow so long that they get in the way. A good pruning schedule should be followed even if the canes are allowed to sprawl on the ground without being tied to supports, or the boysenberry patch will soon be an unmanageable sea of vines and thorns. In no event should more than 6 to 12 canes be allowed to grow from one plant.

Pests and Diseases

Boysenberries are subject to the same insect enemies and diseases as raspberries (see Index). Birds particularly enjoy boysenberries but can be foiled by netting the berries as they ripen or using any of the measures described in Appendix I, For the Birds. Boysenberries will attract few insects and suffer fewer diseases if grown on a trellis or wires. Practicing sanitary measures in the garden is also very important—pruned wood should be burned and the number of vines regulated. Never plant boysenberries on soil recently used for tomatoes, peppers, or eggplant, either, since these crops sometimes transmit a soil-borne wilt disease to the plants.

Propagation

Dividing old boysenberry plants into new ones is hardly practical. Tip-rooting, the way that boysenberries multiply themselves in nature, is the best way to increase the cultivated boysenberry, too. In midsummer cover the cane tips with two inches of soil and leave in place until the next spring.

During this time, roots and stem buds will develop under the soil and after they are cut off the mother canes they can be transplanted to their permanent place in the garden. It is much easier to start new plants like this than to move established bushes, for old berry plants don't transplant easily. But if you choose to move established plants, the best time to do so is in early spring before growth starts. In mild climates, boysenberry plants may be moved any time during the winter.

Harvesting Boysenberries

When boysenberries turn from red to purplish-black, they are ripe. Long fruit spurs growing well out from the stems make the picking of boysenberries fairly easy, but if you are gathering the thorny varieties, take no chances and wear gloves anyway. Harvest the berries carefully as soon as they are dead ripe (when they are as sweet as any berry) and drop off at the slightest touch, for they have a bitter taste when unripe. Don't let them sit too long after picking, either. The berries are very delicate and spoil easily, which is the reason they are rarely carried in stores and must be grown at home to be enjoyed.

Boysenberry Recipes

Handle boysenberries carefully when washing them to prevent their delicate skins from breaking and juice from being lost. A good idea is to harden the berries in the refrigerator for a few hours before cleaning them in ice water. Do not keep them for days in the refrigerator, however, as they mold quickly. The berries are sweet and delicious served as a dessert just as they are after picking, or sprinkled with a little sugar, but they can be frozen, canned, made into a striking violet-red boysenberry wine, or jelly, or a mouth-watering boysenberry pie, and substituted for raspberries in many recipes.

Boysenberry Ice

Bring 2½ cups of boysenberries to a boil in 2 cups of water. Add a pound of sugar and the juice of two lemons, stirring until dissolved. Let stand in a warm place 1 to 2 hours, then rub through a sieve. Add 2 more cups water, mix well, chill, and freeze.

Steamed Boysenberry Dumplings

Substitute boysenberries for the raspberries originally used in this delicious early 1800s recipe: "Cut 4 teaspoons of butter into 2 cups of flour sifted with 1 tablespoon of baking powder and 1 teaspoon salt. Add ¾ cup of milk as you work. Into a buttered baking dish put one quart of berries and mix in 2 cups sugar; then stir in 2 teaspoons of vinegar. Roll the paste into a suitable size to cover the berries. Put it into place and steam, tightly covered, for 1 hour in steadily boiling water that stands high enough along the sides of the dish to cook the dumpling without reaching above the top of the dish."

Boysenberry Pudding

Blend 1 cup of sifted flour into 3 cups of fresh boysenberries. To 2 cups of flour add salt sufficient to season and 1 even teaspoon of baking soda dissolved in 1 tablespoon of milk, adding ¾ cup of maple syrup, and stirring all into a smooth batter. Lastly add the berries, mixing lightly so as not to break the fruit. Put the mixture into a buttered mold and place it in boiling water that does not quite reach the top of the mold. Boil for at least 2 hours.

Boysenberry Delight

1 pint boysenberries
15 graham crackers, crushed
¼ cup melted butter
½ pound marshmallows
½ cup milk
1 cup heavy cream, whipped
2 tablespoons cornstarch
¼ cup sugar
1 tablespoon lemon juice

Crush boysenberries and reserve juice. Combine cracker crumbs and butter and press into a pan. Melt marshmallows in pan over boiling water, cool, add milk, and then fold in whipped cream. In another pan mix juice from boysenberries, cornstarch, sugar, and lemon juice. Cook until thickened and add boysenberries. Pour half of the marshmallow mixture over crumbs. Cover with boysenberry mixture, then top with remaining marshmallow mixture. Chill before serving.

9

Dewberry Delights—
Including Loganberries, Nectarberries, and Youngberries

Dewberries, which are trailing or procumbent blackberries like boysenberries, are a purely native American fruit, although they are cultivated today by gardeners in other parts of the world. Not that there aren't native European varieties of dewberries. In Europe the dewberry can be traced back to the sixteenth century when Shakespeare praised it. Its name, in fact, may be a corruption of "dove berry," which it has been called for centuries in Germany, but it has long been associated with "dew" in English use. English dewberries (*Rubus caesius*) are similar to the American species, though they aren't as tasty. The American dewberry is sometimes dismissed as "just another trailing blackberry," but this is not the case. Dewberries have a better, milder flavor than blackberries, in the opinion of many gourmets, are larger than blackberries, and ripen earlier than the better-known fruit—usually one to two weeks earlier, right in between the raspberry crop and the blackberry crop. A worthwhile addition to any fruit garden, they are easy to grow and can be used in a wealth of recipes ranging from wine to pies.

Dewberry Species And Varieties

Four major species of dewberry are the ancestors of the natural and artificial hybrids grown in American gardens today. All of them grow in the wild and are sought out every spring by thousands of berry gatherers.

Rubus trivialis. Called the *southern dewberry* or the *common dewberry,* this species is found in the South and has black, delicious fruit that can be dry and seedy. Because its canes are very long and trailing it is sometimes called the *running blackberry.*

Rubus canadensis. The *American dewberry* or the *Canadian blackberry,* depending on which expert you consult. The species is native to the mountains of Canada. Its black fruits are large, juicy, and a bit acid. It has long, weak, thornless stems that are semi-erect.

RUBUS VITIFOLIUS. The *western dewberry,* which has been used to develop many hybrids. A California native that prefers the moist soil beside streams, it has sweet black fruit of medium size, borne on weak, slender stems that are usually trailing but sometimes erect.

RUBUS PROCUMBENS. The *field dewberry* bears large, juicy, jet-black fruits on long prostrate canes. It will grow in poorer, drier soil than most dewberries and has been used to hybridize a number of varieties.

The above species can all be collected in the wild and grown in the home garden, or well-tried named varieties of them can be purchased from nurseries. Named varieties include the following:

LUCRETIA. Sometimes called the *bingleberry* and generally considered to be the best cultivated dewberry—the hardiest, the largest, and the sweetest. Its coreless, bluish-black berries often reach 1½ inches long and its canes have a trailing habit. Ripens about 7 to 10 days before blackberries. Lucretia is hardy in zones 7 and 8 and is available from Spring Hill Nurseries, Gurney's, and Field's.

MAYES. An old variety that isn't as hardy or coreless as Lucretia, but to some has a better flavor. Ripens early, the vigorous plants producing beds of large, soft berries. The plants are susceptible to anthracnose and rosette. Hardy in zones 7 and 8, though it needs winter protection in zone 7.

THORNLESS GARDENER. Not only is this Thornless easier to pick than other dewberries; it is practically disease-free as well, and ripens two weeks before blackberries. Crops, however, aren't as bountiful as Lucretia's. Offered by the Earl May Seed & Nursery Co.

YOUNGBERRY. Sometimes this is called the *Young dewberry* or the lavaca. It is actually a hybrid variety of dewberry that ripens before most dewberries and its large, dark-purplish, sweet fruit has the high aroma and flavor of the loganberry and the native blackberry. The youngberry was developed by Louisiana horticulturist B. M. Young in about 1900 by crossing a southern dewberry and trailing blackberry, or by crossing several varieties of blackberries, or by crossing a western dewberry and southern dewberry (no one is certain). Its long, trailing stems are relatively thornless. The berry is extensively planted in the American Southwest, South, Pacific Northwest, and throughout California. It is hardy in zones 6, 7, and 8, but needs to be winter-protected in zone 6. A thornless variety is available from R. H. Shumway, Hastings, and Boatman's.

NECTARBERRY. Because it originated from a youngberry seed and the other parent can't be named, the full ancestry of this rather hardy berry can't be ascertained. Nectarberries are a deep wine color, are borne on long trailing canes, and have a few seeds. They ripen later than dewberries.

OLALLI BERRY. A cross between the loganberry and blackberry very popular on the West Coast. Firm, tender, not grossly seeded berries that are delicious with sugar and cream.

PHENOMENAL BERRY. Much used for making jams, the phenomenal berry is a child of the red raspberry and western dewberry. It bears bright red berries on trailing canes and ripens later than loganberries.

LOGANBERRY. California Judge James Harvey Logan (1841–1921), who had been a Missouri schoolteacher before working his way west as the driver of an ox team, developed the loganberry in his experimental home orchard at Santa Cruz. Logan, formerly Santa Cruz district attorney, was serving on the Superior Court bench in 1880 when he raised the new berry from seed, breeding several generations of plants to do so. Though a respected amateur horticulturist, he never adequately explained how the berry was developed. One account claims that the loganberry originated from "self grown seeds of the Aughinbaugh (wild blackberry) . . . the other parent supposed to be a raspberry, of the Red Antwerp type." Other experts believe that it is a variety of the western dewberry, or a hybrid of that species, crossed with the red raspberry. The dispute may never be resolved, but experiments in England have produced a plant similar to the loganberry by crossing certain blackberries and red raspberries. In any case, there is no doubt that the tart, purplish-red loganberry is shaped like a blackberry, colored like a raspberry, and combines the flavor of both—or that it was first grown by Judge Logan and named for him when the University of California released the berry to the public in 1893. Its scientific name is *Rubus loganbaccus* and the trailing plant is grown commercially in large quantities, especially in California, Oregon, Washington, and other places having fairly mild winters. The delicious berry is hardy in zones 7 and 8. Gurney's offers a new thornless variety.

LOWBERRY. A hybrid of the western dewberry and the Texas blackberry that is sometimes called the black loganberry. The canes trail for up to twenty feet and require trellising. Like the lactonberry it is self-sterile and needs blackberries or loganberries nearby. Its large, black berries taste something like blackberries, not being as acid as loganberries.

WHEN, WHERE, AND HOW TO GROW

The culture of dewberries is identical to that for boysenberries (see Index). Like the boysenberry, the dewberry is more at home in warmer climates, but can also be grown successfully in the North if protected properly in winter. Planting, feeding, watering, weeding, pruning, propagating, harvesting, and protecting against insects and diseases are the same as for boysenberries. Dewberries are best grown on supports, too, using the methods found under boysenberries (see Index), and the black varieties tolerate poor soils (that is, produce a good crop on poor soils) even better than boysenberries.

Recipes

Remember that *all* dewberries can be used in the following recipes and that dewberries, like boysenberries, can be substituted in all the bramble-fruit (raspberry, blackberry, etc.) recipes in this book.

Dewberry Meringue Pie

⅓ cup sugar
2 tablespoons cornstarch
¼ teaspoon salt
¼ teaspoon cinnamon
2½ cups stewed dewberries
2 tablespoons lemon juice
1 tablespoon butter
1 8-inch baked pastry shell
2 egg whites
4 tablespoons sugar

Preheat oven to 350°. Combine the ⅓ cup sugar, cornstarch, salt, and cinnamon with the syrup drained from the dewberries. Cook in a double boiler until smooth and thick, stirring constantly. Continue cooking for 10 minutes. Remove from heat and add the lemon juice, butter, and berries. Pour into the pastry shell. Beat the egg whites until stiff. Add the sugar gradually to the egg whites and beat until mixture will stand in peaks. Spread over pie and bake in a moderate oven about 15 minutes or until well browned.

Loganberry Sherbet

2 teaspoons gelatin
¾ cup sugar
¼ teaspoon salt
1 teaspoon grated orange rind
¼ cup orange juice
12 ounces loganberry juice
½ cup whipped cream

Soak gelatin in ¼ cup cold water 5 minutes. Bring sugar, salt, and ½ cup water to boiling point. Remove from heat. Add gelatin, orange rind, and orange juice. Stir until gelatin is dissolved. Add loganberry juice. Chill. Start freezing. When mushy, remove and beat. Fold in cream. Finish freezing.

Loganberry Chiffon Pie

1 cup graham cracker crumbs
3 tablespoons melted butter
1 tablespoon gelatin
12 ounces loganberry juice
1 cup sugar
¼ teaspoon salt
2 tablespoons lemon juice
1 teaspoon grated lemon rind
1 cup heavy cream
1 stiffly beaten egg white

Mix crumbs and butter and line a 9-inch pie pan. Soak gelatin in ½ cup cold water 5 minutes. Dissolve over hot water. Add to the fruit juice and sugar. Stir until sugar is dissolved. Add salt, lemon juice, and rind. Whip the cream. Fold half of cream and egg white into mixture. Turn into crumb-lined pan. Chill until firm. Top with remaining cream. Sprinkle a few graham cracker crumbs on top.

10

The Regal Raspberry

Because the delicate, delectable raspberry is the hardest of all small fruits to pick, pack, ship, keep, and display, it is rarely found even in gourmet groceries on a regular basis, making it almost essential for those who want an adequate supply of this regal berry to grow their own. Back in 1899, when farm country was closer to the cities, nearly 115 million pounds of raspberries were grown in America annually, but today our yearly harvest isn't half that and most of what is grown is frozen for the supermarkets. Unless you are willing to pay up to $1.50 a half pint *in season* for fresh raspberries (I've seen them at $4.50 a half pint in February) at fancy gourmet food shops, you'll have to harvest your own crop of this highly perishable fruit, which will taste even better picked just before eating anyway.

Fresh raspberries are so good that most people wouldn't dream of wasting the fruits by cooking them (frozen raspberries can be used for that). To my taste they don't even need cream or sugar. Not as celebrated as strawberries, they are more prized for their comparative rarity and are an even better "pin money" crop for those with a large piece of property who want to grow them for the local market. In the North they are second only to the strawberry in popularity among small fruits. The berries aren't the easiest crop to grow, but they involve much less work than many fruits, including strawberries.

Raspberries, which grow wild throughout most of Europe, Asia, and North America, haven't been cultivated as long as blackberries, though they have been the pride of gardeners for over two thousand years. The Roman poet Propertius writes about the raspberry, and the Roman historian Pliny mentions it in his agricultural writings, stating that it had been introduced into Rome from Mount Ida in Greece. Known to the Romans as the "Red Berry of Mount Ida" (hence the name of the British species *Rubus idaeus*), it was and still is cultivated in Europe, especially in the northern countries.

Surprisingly enough, the best French specimen is found within the Paris city limits.

Long called a brambleberry, or "hindberry," and considered a nuisance in England, the raspberry may take its name from the English *rasp,* to scrape roughly, in reference to the thorned canes bearing the berries, but no one knows for sure. First called the raspis-berry, it can be distinguished from its close relative the blackberry by its more delicate flavor and the fact that it easily separates from its core when ripe, while the core of the blackberry is picked and eaten with the fruit. A highly nutritious berry, it has a carbohydrate content of 12 to 13 per cent and is high in vitamins A (25 mg. for 3½ ounces), C (30 mg. for 2½ ounces), B_1, B_2, and contains calcium, magnesium, phosphorous, and iron salts. Just the opposite of the blackberry, it is considered a good laxative and diuretic and because it is rich in pectin is one of the best fruits for jams and jellies. Besides being eaten fresh, or with sugar or cream, raspberries are used to make delicious charlottes, summer puddings, pies, tarts, raspberry cream, raspberry fool, ice cream, sherbet, raspberry-ades, wine, and even vinegar. One delicious old American drink is raspberry shrub, which has been enjoyed here at least since the early nineteenth century.

It was not until about 1830 that the regal raspberry began to be developed in America, wild plants usually being plowed under to clear farm land before that. Hybridists, including Luther Burbank, have since worked with the native American, the European, and the Asian raspberry to develop remarkable new varieties that are delicious, large, prolific, long-bearing, and easy to grow. Of all the sweet, multicolored, everbearing, even thornless varieties that have been introduced, probably the nicest story is told about the Fanny Heath raspberry. This variety is a tribute to a determined pioneer woman who emigrated to North Dakota in 1881. The young bride had been warned that she could never grow anything in the barren alkaline soil surrounding her house, but forty years later her homestead was an Eden of flowers, fruits, and vegetables. After her death in 1931, the black raspberry she developed was named in her honor.

Gathering Wild Raspberries

As we'll see, red, black, purple, yellow, thornless, and everbearing varieties of raspberries are offered by nurseries, and the French even claim a white raspberry, a type of the red variety Merveille des Quatre-Saisons (see European Red Raspberries, in the Index). But as an additional bonus there are a number of kinds that can be found growing in the wild. In fact, almost all commercial varieties listed later come from patient crosses of the following raspberry species, some of which are native to North America.

Rubus deliciosus. The *Boulder raspberry* or *Rocky Mountain flowering raspberry*. Despite its scientific name, the dark reddish-purple fruit on this species is almost worthless. The plant, however, is thornless and shrubby, valuable for breeding work. Hardy from zone 4 southward.

RUBUS ELLIPTICUS. The *Himalayan golden raspberry.* A tall, upright plant with reddish-brown hairs on its stems and good-quality yellow fruits of average size.

RUBUS GLAUCUS. The *Andes raspberry.* A black raspberry native to South America.

RUBUS IDAEUS. The *European raspberry.* An erect shrubby plant with bristly stems that may have originated in Asia but is naturalized in Europe, and bears average-sized, thimble-shaped, red berries. This is the Red Berry of Mount Ida that Pliny extolled and the ancient Greeks and Romans enjoyed.

RUBUS ILLECEBROSUS. The strawberry-raspberry (see Unusual Fruits, in the Index).

RUBUS LEUCODERMIS. The *Western black raspberry.* Native to western North America.

RUBUS NIVENS. The *Asian raspberry.* A species native to Asia, bearing black fruit.

RUBUS OCCIDENTALIS. The *North American blackcap raspberry* or *thimbleberry.* An erect, prickly plant with bluish stems that root at the tip, creating impenetrable thickets. Its fruit is black with a slight bloom, and it is hardy from zone 2 southward.

RUBUS ODORATUS. The *flowering raspberry.* A showy shrub with sticky but not thorny stems and red, dryish fruit that is of very poor quality if not worthless. Hardy from zone 2 southward.

RUBUS STRIGOSUS. The *North American red raspberry.* This plant is very similar to the European red raspberry (*Rubus idaeus*), but yields a flatter fruit. It may, in fact, be a variety of *Rubus idaeus,* but, in any case, is the source of most present-day red raspberry varieties.

THE BEST RASPBERRIES TO PLANT

It should be noted that some authorities estimate there are as many as a thousand species of the brambles, most of which are not cultivated but can be found growing in the wild. Nevertheless, for the most part the ten preceding species were crossed and recrossed to obtain the red, yellow, black, purple, thornless, and everbearing varieties so popular today. As we'll see, the closely related reds and yellows are generally cultivated in the same way, and somewhat differently than the more closely related blacks and purples. Generally speaking, raspberries grow best in cool climates—and there are many excellent varieties for cold sections of the country. But there are varieties for hot, humid climates, too, Southland (an everbearer) probably being the best. The time lapse between ripening of early and late varieties listed here may be as little as twenty days and as much as ninety days or more. Choosing these varieties wisely and utilizing the everbearers, one can harvest raspberries from June to October in many sections of the country. The general rule of thumb is that twelve plants of any of these varieties should be enough for a family of four, unless a lot of preserving is planned or you intend to sell berries (don't forget that the plants will increase). But you can figure it for yourself—properly tended raspberries can produce from 2,000 to 4,000 quarts an acre, or about one quart of fruit for each plant. The plants usually bear a full crop 1 to 3 years from planting, and flowers should be pinched off them the first year to strengthen the plants.* Where a berry is particularly difficult to obtain, a source is given here in brackets at the end of the entry.

Red Raspberry Varieties

CANBY. A midseason berry developed in Maryland by the USDA that isn't adapted to heavy soils but is one of the top varieties in the Pacific Northwest for freezing. The berries are large, firm, light, bright red, and sweet, the plants vigorous, hardy, productive, and practically thornless for easy handling.

CHIEF. A berry developed in Minnesota and the standard early variety for the upper Mississippi Valley. Bushes are vigorous, productive, and among the hardiest; the berries are smallish, bright red, firm, and of good quality.

CITADEL. Developed in Maryland for the Middle Atlantic states, this variety bears medium-sized, bright-red berries on a very hardy, productive plant.

CUTHBERT. An old favorite, yielding bright red medium-sized fruit of good quality that is moderately susceptible to disease and has been largely replaced by Latham.

FAIRVIEW. An Oregon breed, grown there and in Washington, this midseason variety succeeds on heavy soils and is resistant to root rot. Berries are large, bright red, firm, of very good flavor, and good for freezing. Plants are vigorous, productive, and have long-fruiting laterals.

HILTON. Developed in New York for the northeastern United States, this midseason variety is the largest of the red raspberries. The berries are very attractive—firm, medium red, darkening quickly—but are hard to pick. The plants are productive with stiff, erect canes [Spring Hill].

JUNE (also called *Ontario*). A very early variety with a long season that was developed in New York, is quite hardy in the East, and is adapted

* Large, bearing-age, transplanted bushes of some varieties are available for those who want some fruit the first year and are willing to pay more. Sources include Field's, Shumway's, Spring Hill, and Gurney's. Addresses for all these suppliers are listed in Appendix III.

to heavy soils in New York, New England, Michigan, and Wisconsin. The berries are large, bright red, firm, and keep better than most, although they sometimes lack good dessert quality. The bushes are vigorous, almost thornless, comparatively free from disease, but sucker rather poorly.

LATHAM. An old, very popular late variety bred in Minnesota and one of the hardiest, hardy even in North Dakota. Large, roundish, medium red, moderately firm berries that are inclined to crumble and aren't of the highest dessert quality but are good for canning and freezing. Plants are usually vigorous, very productive, and nearly thornless, but susceptible to mildew. Useful primarily where better varieties winterkill; especially tolerant of the mild winter weather prevailing south of New York City that causes much winter injury.

MADAWASKA. Developed in Ontario, Canada, this early berry is very winter-hardy and well adapted to the New England states. The plants are productive, the berries of fair size and quality but somewhat acid and dark.

MILTON (*New Milton*). A late variety that is a New York development and, because of its high resistance, is grown in the East, where mosaic viruses are a problem. The berries are large, firm, conical rather than round, and of good flavor. The plants are tall, vigorous, and moderately hardy.

NEWBURGH. A midseason New York-bred plant slightly earlier than Latham that is often grown in the Northwestern states and the Pacific Northwest. Berries very large, bright red, firm-fleshed, mild, and of good quality. Plant vigorous, hardy, very productive, and nearly immune to diseases—including mosaic and root rot. A standard and reliable variety.

ONTARIO (see *June*).

ORNAMENTAL RASPBERRY. The best raspberry for making practically impenetrable 4 to 5-foot hedges. Bears early and yields berries enclosed in red-orange prickles [Burgess].

OTT'S PENNRIDGE HILL. A midseason variety with good-quality fruit that does especially well in the Northwest.

PUYALLUP. A Washington development, this late variety is much grown in the Pacific Northwest but isn't adapted to heavy soils. The berries are large, bright red, somewhat soft, have very good flavor, and are of good quality when frozen or canned. Plants are hardy, vigorous, and moderately productive.

RANERE (*St. Regis*). Another old favorite that is early fruiting, though later than June, and often fruits again in late summer. Fruit not as good as Latham or June.

REVEILLE. A Maryland development that is more hardy in the Middle Atlantic states than varieties of northern origin. Medium-sized, good-quality fruit on a productive plant.

SCEPTER. Also developed in Maryland and similar to Reveille, above.

ST. REGIS (see *Ranere*).

SUMMER. A late variety developed in Washington that is well adapted to the heavy soils of the Pacific Northwest, is cold-hardy and resistant to root rot and yellow rust. Berries are medium-sized, firm with high flavor. Plants are vigorous, hardy, and productive.

SUNRISE. Developed in Maryland, this very early variety (perhaps the earliest of all) is widely grown in the northeastern and eastern north-central United States. Medium-sized, firm berries with good flavor. Vigorous plants that sucker freely and have large basal fruiting shoots [Buntings' Nurseries, Inc.].

TAYLOR. A midseason to late variety developed in New York that is a leading seller there and in New England and is well adapted to all northeastern states. Berries conical, very large, firm, of high quality, and excellent for freezing. Plants vigorous, sturdy, and erect, usually not needing support, and very productive.

VAN FLEET. This old midseason variety bears only moderately well and its fruit doesn't compare to the high-quality raspberries, but it does well in warm areas unfit for most varieties.

VIKING. An old type bred in Ontario, Canada, that is still grown in New England and the North Central states. Berries medium-sized, firm, of excellent quality. Canes vigorous and spineless.

WASHINGTON. A Washington introduction that bears fruit late and is a major variety in the Pacific Northwest for canning and freezing. Berries large and of high quality. Plants vigorous and productive.

WILLAMETTE. A midseason variety from Oregon that is also grown extensively in the Pacific Northwest for freezing and canning. Berries very large, nearly round, medium red, very firm, of good quality. Bushes vigorous, productive, sucker freely.

European Red Raspberries†

BAUMFORTH'S SEEDLING. Fruit large, dark crimson, of excellent flavor. Plants vigorous, productive, and bear in midseason.

BELLE DE FONTENAY. A late variety with large round fruit on productive plants.

CARTER'S PROLIFIC. A midseason variety with large, deep-red, firm fruit that bears very freely.

FALLRED (see Everbearing Raspberry Varieties, in the Index).

FASTOLF. Very large fruit of good flavor on this abundant midseason bearer makes it one of the most generally useful sorts.

LLOYD GEORGE. Probably the best-known English red; an early English variety named for the famous Prime Minister. Berries very large, dark red, of excellent quality. Like most English raspberries, plants are only medium in hardiness, vigor, and productiveness. The American red Canby (see Index) is a cross of Viking and Lloyd George.

PRINCE OF WALES. Bears large, delicious fruits in midsummer on plants with unusually strong canes, but is not prolific.

RED ANTWERP. A very old midseason variety that is much cultivated, with large, conical berries of good flavor.

† European reds are generally considered better in quality, but aren't as hardy as American plants. The above are all English varieties, but the French also have developed some excellent raspberries, including Perpétuelle de Billard, Merveille des Quatre-Saisons, Horney, and Pilate. Only overseas nurseries have a supply of English and French varieties. (Try Sutton for English types, Vilmorin for French kinds—see Appendix III.)

Yellow Raspberry Varieties
(a mutation of red raspberries)

AMBER. A midseason variety with long, conical berries amber in color that are tender and sweet, of very good quality. Plants very tall, vigorous, hardy, and productive after getting established [New York State Fruit Testing Cooperative].

FALLGOLD (see Everbearing Raspberry Varieties, in the Index).

GOLDEN MAYBERRY. One of the earliest raspberries, ripening before strawberries, this variety has large, sweet, juicy, golden berries. Plants are 6 to 8 feet high with a spreading top, the result of a cross between a Japanese yellow and red by Luther Burbank. Burbank's account of the plant's history gives some insight into plant breeding. "Some ten years ago," he wrote, "I instructed my collector in Japan to hunt up the best wild raspberries, blackberries and strawberries that could be found. Several curious species were received the next season and among them a red and also a dingy yellow unproductive variety of *Rubus palmatus.* One of these plants, though bearing only a few of the most worthless, tasteless yellow berries I have ever seen, was selected solely on account of its unusual earliness to cross with well-known raspberries. Among the seedlings raised from the plant was this one [the Golden Mayberry] and though no signs of the parent appear it can hardly be doubted that the pollen of the other parent effected some of the wonderful improvement to be seen in this new variety."

GOLDEN QUEEN. Midseason, medium-sized, sweet golden berries on very hardy plants.

GOLDEN WEST. A new midseason variety developed in Washington for the Pacific Northwest. Large, sweet, golden berries on a productive plant that is somewhat disease-resistant.

Black Raspberry Varieties (Blackcaps)

ALLEN. A New York-bred midseason variety (a cross of Cumberland and Bristol) with large, firm, attractive fruits that ripen many at a time, and are much used for jam making in the North Central and Northeastern states. Said to be the sweetest of the blackcaps. Bushes are vigorous and productive.

BLACK HAWK. Plant breeders at Iowa State College tested this new very late variety twenty years before releasing it and it is said to outbear all other blackcaps (5,000 pints per acre). Widely grown in the Midwest and the eastern United States, the berries are large, firm, and glossy, with good flavor. Plants are vigorous, productive, very hardy, but do not sucker. Somewhat anthracnose-resistant, they are susceptible to mildew [Field's].

BLACK PEARL. A Missouri early-ripening variety grown chiefly in Kansas and Missouri. Berries large, firm, glossy, of good quality. Plants drought-resistant and nearly immune to disease.

BRISTOL. Developed in New York, this midseason variety is widely grown in the eastern United States. Berries large, conical, firm, high-flavored, and almost seedless, but hard to pick after rain. Plants tall, vigorous, very productive, and generally disease-resistant, but very susceptible to anthracnose.

CUMBERLAND. No black raspberry has been grown as extensively in all sections of the country as this Pennsylvania midseason variety, which is best for jams and pies. Its large, firm, conical berries are said to be the best of all blackcaps in flavor. The plants are vigorous and productive, but susceptible to many diseases. New Improved Cumberland ripens two weeks earlier and has more disease resistance.

DUNDEE. A midseason New York-bred variety with large, firm, glossy, good-flavored berries that are hard to pick after rain. Plants are vigorous, hardy, and productive, but susceptible to mildew.

FARMER (see *Plum Farmer*).

HURON. A late midseason variety developed for western New York that yields large, glossy, firm berries of good quality. Bushes hardy, vigorous, productive, and somewhat anthracnose-resistant.

JEWEL. A New York-bred midseason variety with large, firm, glossy berries on vigorous, productive plants.

LOGAN (see *New Logan*).

MORRISON. An Ohio-bred late variety primarily grown there and in New York and Pennsylvania. Berries are firm, glossy, of fair quality, and probably the largest of the blackcaps. Bushes productive. New Morrison is an improvement with berries an inch in diameter [Field's].

MUNGER. Another midseason Ohio berry that is a leading variety in Oregon. Berries large, firm, of good flavor, but plants susceptible to mildew.

NAPLES. An old late variety still grown that yields large, glossy, slightly acid fruit on a hardy plant nearly immune to disease.

NEW LOGAN (*Logan*). An Illinois midseason variety ripening a week before Cumberland that conveniently ripens its crop almost all at the same time, making for easier picking. Widely grown in the eastern United States and Michigan. Berries medium-sized, of good quality; plants very productive.

PEARL (see *Black Pearl*).

PLUM FARMER (*Farmer*). This early-season Ohio-bred berry ripens so quickly that the entire crop can be harvested in 2 to 3 pickings. The berries are large, firm, of high quality, but have a slight bloom. The bushes are hardier than most blackcaps and drought-resistant. An important variety in Oregon, susceptible to anthracnose but immune to curl virus.

SHUTTLEWORTH. A New York development that bears an early crop of medium-sized, glossy, firm berries of good quality. Plants vigorous, productive, and moderately disease-resistant.

Purple Raspberry Varieties
(crosses between red and black varieties)

AMETHYST. Developed in Iowa and tested over a twelve-year period, this is a good midseason variety for the Midwest. Berries are large, shiny, of good quality, and always visible on the stem. Plants are very hardy, productive, and resistant to disease.

CLYDE. A late New York variety that is grown mainly in the northwestern United States. Berries large, firm, and tart. Bushes hardy, very productive, with stout, vigorous canes, and moderately anthracnose-resistant [J. E. Miller].

COLUMBIAN. An old-fashioned late variety with large, dark-purple fruits of good quality. Plants prolific but quite susceptible to disease.

MARION. Another New York late variety popular in the Northeast. Large, firm, tart berries of good quality on vigorous, productive bushes.

PURPLE AUTUMN (see Best Everbearing Varieties, in the Index).

PURPLE SUCCESS. A new midseason variety that bears large, firm berries with a pleasant, tart flavor. Plants vigorous, upright, hardy, productive, and disease-resistant.

SHAFFER. An old variety very similar to Columbian, fruits not quite as good, but more productive and less susceptible to disease.

SODUS. Developed in New York, this midseason variety is widely planted in the Northeast. Berries are very large, fairly firm, not crumbling, and of good quality though quite tart. Plants are vigorous, productive, drought-resistant, winter-hardy, and very disease-resistant. Plants have the growth habit of black raspberries, and are propagated by tip-layering.

Everbearing Raspberry Varieties

(Two-season berries that bear once in spring, early or late summer on last year's canes, and again in autumn on canes of the current season's growth.)

AUGUSTRED. A new variety that yields some fruit early but bears its real bumper crop in late August. Large, sweet, bright-red berries on erect, productive plants particularly adapted to the North.

BLACK TREASURE. An everbearer that fruits from June until frost, yielding large, delicious berries.

DURHAM. Introduced by the University of New Hampshire in 1942 this red everbearing variety is extremely hardy and has good flavor. Bears a heavy crop in early summer on old canes and produces again in a few weeks on new canes. New Durham is an improvement with larger, juicier berries, free from seediness.

FALLGOLD. Similar to Fallred, below, but with golden fruit and even better flavor, many experts considering it the best-tasting raspberry.

FALLRED. Developed for the northeastern United States, this excellent variety is productive, vigorous, and hardy, yielding a late summer and fall crop of the biggest berries you can grow; berries have good flavor, too.

HERITAGE. The firmest red berries of high quality on a plant with strong, upright canes that bears its first crop in July and its second from September until the first hard frost. Developed in New York, Heritage is one of the best everbearers for the East.

INDIAN SUMMER. Many consider this to be the best-flavored of the red everbearers; its one defect is that the large berries are crumbly. Plants are vigorous, hardy, and disease-resistant, producing a first crop in July and a second from September to the first hard frost. Good for all sections from South Carolina to Canada [Hastings].

PURPLE AUTUMN. The best of the purple everbearers. Large, sweet berries on a vigorous, productive plant.

REGENT. An old red variety now usually replaced by Indian Summer.

SEPTEMBER. A New York-bred red variety grown extensively in the eastern United States that bears its first crop in July and its second in mid-fall. Medium to large firm berries that have a good, tart flavor, if that's what you're looking for. The plants are vigorous, only moderately productive, hardy, and disease-resistant.

SOUTHLAND. A new variety developed at North Carolina State University that has done well in tests from North Carolina to Arkansas, extending the range of the raspberry, which doesn't normally thrive in warm, humid areas. Produces its first crop very early in June and its second in mid-August. Definitely the best berry for the South [J. E. Miller].

Outstanding Raspberries‡

As a recap here are the best of the 75 or so varieties covered:

BEST-FLAVORED OF ALL RASPBERRIES— Fallgold (everbearer)

LARGEST OF ALL RASPBERRIES— Fallred (everbearer)
BEST-FLAVORED RED RASPBERRIES— Taylor; Fairview; Viking
LARGEST RED— Hilton
MOST PROLIFIC REDS— Latham; Newburgh; Taylor; Willamette
EARLIEST RED— Sunrise
BEST THORNLESS REDS— Canby; June; Viking
BEST-FLAVORED EUROPEAN RED— Lloyd George
BEST-FLAVORED YELLOW RASPBERRIES— Fallgold (everbearer); *Amber*
LARGEST YELLOWS— Golden West; Golden Mayberry
MOST PROLIFIC YELLOWS— Amber; Golden West
EARLIEST YELLOW— Golden Mayberry
BEST-FLAVORED BLACK RASPBERRY— Cumberland
LARGEST BLACK— Morrison
MOST PROLIFIC BLACK— Black Hawk
EARLIEST BLACK— Plum Farmer
BEST-FLAVORED PURPLE RASPBERRIES— Columbian; Amethyst
LARGEST PURPLE— Sodus
MOST PROLIFIC PURPLE— Clyde
EARLIEST PURPLE— Amethyst
BEST-FLAVORED EVERBEARERS— Fallgold (yellow); *Indian Summer* (red); *Fallred*
LARGEST EVERBEARERS— Fallred (red); *Fallgold* (yellow)
MOST PROLIFIC EVERBEARERS— Fallgold (yellow); *Durham* (red)
EARLIEST EVERBEARER— Fallgold (yellow)

‡ Flavor, of course, is a purely subjective matter of taste. As noted in the lists, there are many tart berries some people find delicious. All the colors have their champions, too, including the purples, which don't look as attractive as the others when boxed.

Warning

All raspberries should be grown from guaranteed, virus-free nursery stock certified by your state department of agriculture. This is more costly (perhaps double the price) but worth the expense in the long run, especially when you are buying only a few plants. Virus-free stock is sometimes called "foundation stock" or "registered stock." Remember that such a guarantee only means that the plants were free of virus when you purchased them, not that they can't contract a virus. If the prospect of diseases is your chief concern, buy a certified variety that is described as almost immune to diseases.

Planting Raspberries

Where to Plant

Take care where you place raspberries. Unlike relatively short-lived strawberries, a planting of them can last ten or more years—over twenty years in some cases. Put them off to the side of the garden, or line the small fruit garden with them so that they don't have to be disturbed. Do not plant raspberries near trees that will shade the plants or compete with them for nourishment, and try to eradicate all wild raspberry plants, if any, within a few hundred feet to eliminate the chance of diseases being transmitted to domesticated plants.

Red and yellow raspberry varieties should ideally be planted 600 feet or so away from black and purple varieties because certain diseases carried by reds or yellows (mosaic, for instance) and not usually fatal to them can be transmitted to blacks or purples and kill those types. The farther away the different-colored varieties are from each other, the more difficult it is for aphid disease conveyors to get from one group to the other. The berries, however, will not mix and change color if planted together.

Remember also not to plant raspberries where potatoes, tomatoes, eggplant, peppers, or any member of the potato family has grown for the last three years—for raspberries are susceptible to soil-borne diseases caused by these plants. Never plant these crops close by raspberry rows either.

Aside from these warnings, just select a sunny site for your berries. Raspberries are not choosy about soil, but generally speaking the reds and yellows do best in sandy or gravelly loams, while the blacks and purples prefer heavier, moister loams and can even be planted in clays. The soil's acidity or alkalinity isn't too important, but a pH of from 5.5 to 7.0 is best for all raspberries; any soil below 5.0 should be limed.

Far more important than topsoil conditions for the raspberry is the condition of the subsoil. Like all fruits, raspberries are mostly water and need ample moisture, but they can't stand "wet feet" and must be provided with good drainage. Wet soil not only makes plants "run to wood" at the expense of fruiting, but stimulates autumn green growth that is subject to winterkill. Therefore, since raspberries have root systems that grow three feet and more down, the subsoil should be deep (without a shallow hardpan that cuts the roots off from water) and well drained. The berries require a spot where they will always have enough moisture—never a dry spot—but one where the soil doesn't remain wet all of the time.

Another very important factor is the slope and exposure of the planting site. Where winters are severe raspberries are in less danger of winter injury when planted on hillsides than in valleys. Plants placed on high land will also be less subject to fungus diseases, which thrive in humid areas where air doesn't circulate freely.

Raspberries planted in southern growing areas are also best planted on high ground, but they should additionally occupy sites with a northern or northeastern exposure, which retain humus and moisture longer than sites with southern exposure.

Preparing the Soil

Although raspberries will grow even in poor soil, the best crops are obtained from soil well supplied with organic matter. Rid the planting site of weeds and incorporate plenty of humus into the ground, whether it is strawy, well-rotted manure, compost, leaf mold with a little lime added to it, grass clippings, or a cover crop turned under. Try to prepare the site the season before you set out the plants and dig the humus in to a depth of at least 6 inches.

Cover crops that can be planted and turned under for "green manuring" light soils include rye, oats, barley, vetch, clover; buckwheat can be used where soils are heavy. I have never found it necessary, but some gardeners treat the soil with chlordane dust or spray to control insects like grubs, especially on land that has previously been in sod. This should be done after fertilizing, and carefully, in accord with all directions on the insecticide.

When to Plant

Spring is the safest time to plant all raspberry varieties in most sections, but fall plantings do well in the South, where there is enough time for the raspberries to become established before winter (if there isn't, the plants will be subject to severe winterkill). If you want to try fall planting in the East, do so during the last two weeks in October and, in addition to all other planting precautions, mound several inches of soil around plants to prevent heaving from frost-action, leveling the ground again in spring.

Do not plant at any time when the soil is very wet—especially when it is wet from spring rains. Store plants in a cool place and wait until the soil is dry enough so that it doesn't pack. Otherwise, the new shoots or canes developed from the plant's small underground leader buds will find it difficult pushing up through the ground.

How to Plant: Spacing and Setting Out

If planting must be delayed a considerable time for any reason, protect raspberries from drying out by heeling them in; that is, dig a trench deep enough to accommodate the roots, spread the plant roots down along the trench, and cover them with moist soil. Should the plants have to be kept over winter, pack them in moist sphagnum moss and store at a temperature of 32° F., keeping them moist.

It is good to "puddle" plants, soaking the roots in a mud solution for several hours before heeling them in or planting them—especially if plants are dry when received. While setting out raspberries, always keep them covered with wet burlap or plastic film until they are planted. The point is never to let raspberry roots dry out under any circumstances.

Space red and yellow raspberries 2 to 3 feet apart in rows 5 to 8 feet apart, with no row wider than 1½ feet. Black and purple varieties, more vigorous, should be spaced 5 to 6 feet apart in rows 5 to 10 feet apart. The greater the distance you can space between rows, the easier it will be to cultivate the plants with hand tools or a garden tractor.

Raspberries can also be grown in the "hill system," which has nothing at all to do with hills. Here raspberries are simply set in groups or clumps and can be cultivated on all four sides instead of just on two sides as in the row system. When planting in hills, set red and yellow raspberries 5 to 6 feet

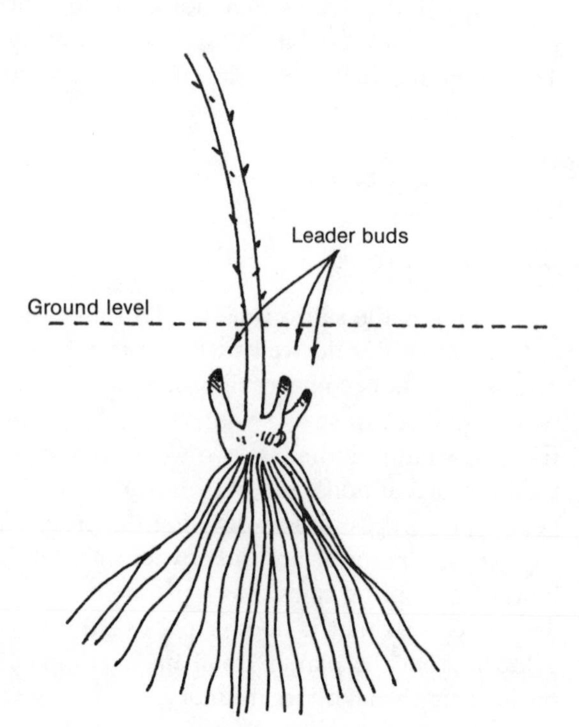

Red and yellow raspberries should be planted 2 to 3 inches deeper than they were in the nursery; black and purple varieties should be no more than an inch deeper than in the nursery.

apart from each other in "hills" or areas 5 to 6 feet apart. Blacks and purples are set 6 to 7 feet apart from each other in hills 6 to 7 feet apart.

Before setting plants cut canes back to about 6 inches long so that you can use the 6-inch top as a handle. Dig a hole about one foot in diameter and work into the bottom a generous amount of compost or other organic fertilizer (never let commercial fertilizer touch a plant's roots). Set red and yellow varieties about 3 inches *deeper* than they grew in the nursery (you can easily tell by the soil mark on the canes), but plant blacks and purples *at the same depth* at which they grew, or no more than one inch deeper.

Spread roots out when planting and firm the soil over them as you fill the planting hole, watering to prevent air pockets.

After planting, the canes of red and yellow raspberries can be left alone, but the canes of black and purple varieties should be cut to the ground and disposed of to help control the spread of diseases on them.

Raspberry plants should be cultivated the first spring they are planted. Do not mulch them until summertime.

Space Savers

For the first year only space can be saved in the garden by intercropping vegetables between the rows of raspberries. Try beans, peas, summer squash, cauliflower, or cabbage. Do not grow potatoes, tomatoes, eggplant, or any members of the potato family, for, as noted, wilt diseases that affect these plants also affect raspberries. Neither should grain crops be grown, for they rob raspberry plants of too much moisture and too many nutrients.

Intercropping shouldn't be practiced after the first year, as bearing-age raspberries need all available soil nutrients and moisture to yield at their best. However, some gardeners do intercrop strawberries and raspberries, planting a strawberry plant between each two raspberry plants in the row. These berries are reported to be very compatible, the strawberries showing better color and growth than when planted alone.

Raspberries will make a nice hedge anywhere on your property if you let them grow 10 to 12 inches apart and don't allow additional plants or suckers to develop in the row. Pull out all other plants, especially those developing at the sides and keep the plants neatly pruned. If an upright variety is used and plants are pruned regularly, no supports will be necessary; if not, wire-strung bean posts can be used.

PRUNING

In order to strengthen your plants, pinch off any blossoms that develop the first year—you won't get a full crop of berries until the second year after planting anyway. Otherwise, do not prune the first year. The second year, pruning should begin on all varieties. Remember that toward the end of each season new raspberry canes of the current season's growth send out laterals (side branches). The following season small branches grow from the buds on these laterals and fruit is borne on these small branches. To obtain bigger and better fruit cut the laterals back in spring before growth starts, leaving six buds per lateral on stout canes and two buds per lateral on thin ones. Canes, of course, will bear no more fruit after fruiting once (except on everbearers) and should be cut out immediately after harvest to prevent possible spread of disease.

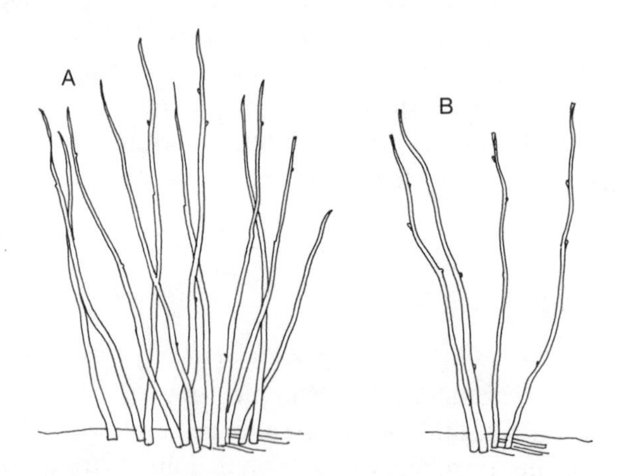

Red raspberry plant before (A) and after (B) pruning.

It is also important to top raspberries every year, especially black and purple raspberries, which double their yield when topped and can be formed into compact bushes instead of sprawling, uncontrolled plants. Topping simply means pruning off the tips from new shoots to prevent them from sprawling and creating a miniature jungle. Topping encourages the canes to branch out. Black raspberry canes should be topped to a height of 18 to 24 inches, purples at 30 inches, and reds and yellows when about 30 inches high. Do not do any topping when the canes are wet or rain is forecast that day.

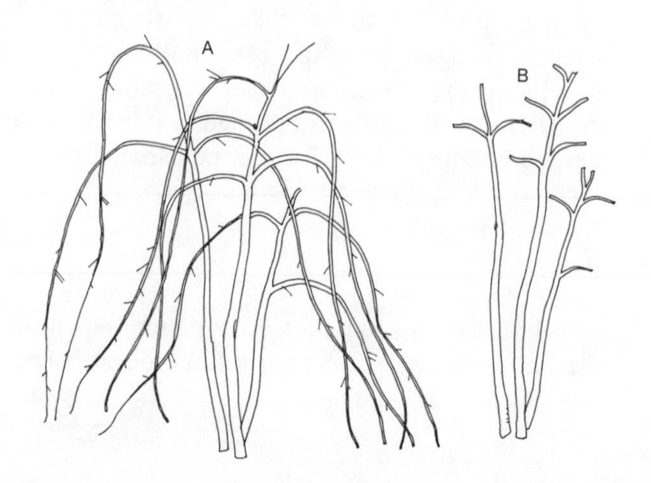

Black raspberry plant before (A) and after (B) pruning.

The only other pruning points to remember are to cut out all diseased and damaged canes and to wear a heavy pair of gloves.

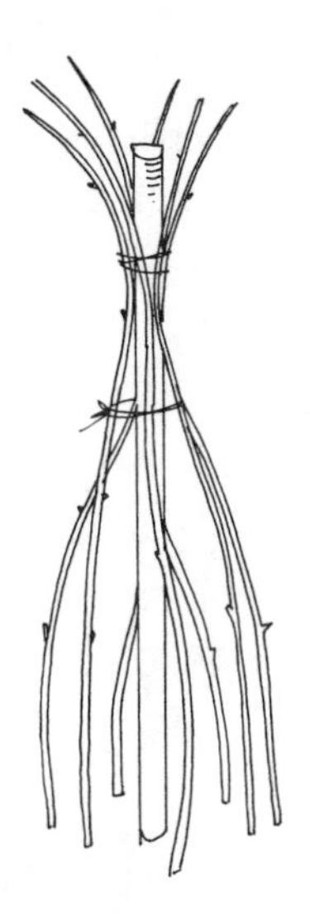

A hill of vigorous red raspberry canes pruned to a height of 4 feet and supported by a stake.

Pruning Everbearing Varieties

"Everbearing" raspberries actually bear berries twice a year: once in spring or summer on canes of the previous year's growth, and again in fall farther down on the same canes. Therefore, everbearers should not be pruned after the fall harvest, since this would remove buds that would fruit the next spring. Canes of everbearers should be cut out after they have fruited in the spring or summer.

Some gardeners prefer to use everbearers to yield a fall crop only, as they produce raspberries later than all other varieties. In this case, the early crop (which is never huge, anyway) is sacrificed. This is accomplished by cutting down all the canes 2 to 3 inches from the ground in early spring and putting all the plant's strength into the development of fall fruiting canes.

Thinning Raspberry Canes

New red and yellow raspberry canes grow from buds on the base of old canes and usually 2 to 3 or more new shoots come up from each old cane every year. Additionally, suckers grow from the roots of red and yellow raspberries. Both new canes and suckers should be thinned immediately after harvest each year (starting the year after planting to prevent crowding of the rows or hills; otherwise the plantation will turn into an impenetrable jungle that doesn't yield well, either). Remove all weak new shoots and most of the suckers, making especially sure to get all those outside the rows and leaving 2 to 3 healthy canes per foot of row, or 6 to 8 plants per hill. Do this every year and be sure to root out suckers, not just cut them to the ground. Healthy suckers that are removed can be used to start new rows or hills.

Black or purple raspberries should be thinned by removing every cane under ½ an inch in diameter. Most black or purple varieties will have 4 to 5 canes over ½ inch, but if all their canes are smaller than this, cut out all but the two largest ones.

Training Raspberries

Raspberries that are pruned and thinned as suggested above usually won't need to be trained to supports, but trained plants are easier to pick, save space, and generally look a lot neater. The same system can be used to train them as described for blackberries (see Index). Black and purple raspberries, which aren't as stout-caned as the reds and yellows, are generally trained more than other varieties.

Raspberries can also be trained by setting two posts a foot apart from each other at each end of each row. Run wires down the row from post to post on each side, arch half the canes from each plant over each wire, and lock them under the canes of the next plant.

Care Of Raspberries

Cultivation

Keep raspberries free of weeds for the first year by cultivating frequently. Begin in early spring and continue until late summer, but *do not* cultivate in the fall—this will only stimulate new growth that will winter-kill easily. Whenever you cultivate be certain to go down only 2 to 3 inches beneath the soil in order not to harm the shallow roots of the plants.

The herbicides Simazine and Diuron can be used according to manufacturer's directions to control weeds that are difficult to reach within rows or hills. Such herbicides should only be used during the dormant season, never while plants are growing. In my opinion, they are totally unnecessary in a small planting that can easily be hoed or mulched.

Fertilizing and Watering

Organic fertilizers are best for raspberries; indeed, some experts say that commercial fertilizer is wasted on them. Big growers often apply 10 tons per acre of stable manure early in the spring just as growth begins, which amounts to about a heaping shovelful of composted manure around each plant. Another good bet is rich compost, which cannot be applied too liberally. If commercial fertilizer *is* used, do not place it too close to the roots or stems of plants, broadcasting it evenly by hand. Choices here include ½ cupful of 5-10-5 fertilizer for each plant; 10-10-10 fertilizer at the rate of 1 pound per 10 feet of row; ammonium nitrate at 1 ounce per plant; or nitrate of soda at 2 ounces per plant.

As for watering, be sure that raspberries get 1 to 2 inches of water (including rain) once a week during the fruiting season and once every 2 to 3 weeks during the rest of the growing season. Light, sandy soils will need even more water.

Mulching

Mulching raspberries is strongly recommended, this method being much superior to cultivating them with a hoe or using herbicides. Not only will a mulch hold down weeds, but it will feed the plants and conserve moisture in the soil. Mulching of raspberries can be started the summer after spring planting. Material to use includes leaves, sawdust, wood chips, straw, hay, crushed corncobs, leaves, and even grass clippings (see Appendix II). Apply the mulch 4 to 6 inches deep, except in the case of hay, 10 inches of which should be used. The best mulch is probably sawdust, wood chips, or straw, but these can cause a decrease in plant growth because they tie up available nitrogen in the soil—so add an organic nitrogen fertilizer such as cottonseed meal, dried blood, or soybean meal to the mulch. If you use inorganic fertilizers, apply a handful of ammonium sulfate for each bushel of sawdust, wood chips, or straw.

Mulch on plants must be thinned out after harvesting to allow new canes to come up straight, but be sure to restore the mulch afterward and to replenish it every year as it decomposes. It's also a good practice to pull mulch away from the plants when spring is cold to let the ground around the plants warm up and the canes leaf out. Mulched raspberry plants will not only grow better but will produce bigger berries and more of them.

Winter Protection

Raspberries need protection from cold, drying, winter winds in certain sections of the country, especially in parts of Colorado and in the western North Central states. Winter injury can be avoided to a large extent by planting the berries on an elevated site, as cold air settles to low areas like valleys and hollows. However, the canes can also be protected winters by bending all of them over in the same direction and holding them close to the ground with clods of earth that are removed in the spring. Other protection methods are covered under blackberries (see Index).

Insect Pests and Diseases

Insect and disease damage to raspberries will be minimal if all the precautions previously noted are taken:

- Plant healthy stock, certified virus-free and disease-resistant, to help prevent mosaic and other diseases.
- Don't plant where diseased plants have been grown.
- Destroy all wild raspberries growing nearby.
- Never plant raspberries where members of the potato family have recently grown and never plant members of the potato family close by raspberries.
- Plant black and purple raspberries at least 600 feet from reds and yellows.
- When planting, cut off old stubs from young raspberry plants.
- Remove old canes after harvest except on everbearers.
- Keep the plantation free of weeds and fallen leaves.
- Destroy diseased plants.

Generally, raspberries are susceptible to the same diseases as blackberries and are damaged by the same insects (see Index). Follow the precautions given for blackberries and you should have little trouble keeping them in bounds. Virus diseases, however, are more destructive in raspberries than in blackberries. Virus diseases can cause poor fruiting, crumbly berries, hard fruits, leaf curl, and yellowing of leaves. While the controls suggested for blackberries can be practiced here, too, the best control is to buy the more expensive "certified virus-free" raspberry plants that are available. This will save you much possible trouble and can't be stressed too strongly.

Following are the most common raspberry symptoms of illness and their causes. *Cures and controls for most of these can be found under blackberries* (see Index), along with other ailments raspberries often share with blackberries.

BLACK-TIPPED CANES. Caused by cane borer or cane maggot.

CANE SPOTS. Anthracnose fungus causes canes to spot, dry up, and die, and causes brown spots on leaves that decay and become holes.

CRUMBLY BERRIES. Anthracnose, mosaic, or winter injury are often responsible, though some varieties have very crumbly berries.

FEW BERRIES. Can be due to many factors, including lack of sun, too much nitrogenous fertilizer, etc., but is frequently caused by winter injury of fruit buds. The control is to cultivate the plants properly, not permit new growth late in the season, to provide good drainage, and to winter-protect the plants as noted previously.

FRUIT ROT. Anthracnose causes rotting of fruit clusters and can be controlled (see Index), but nothing much can be done about fruit rot caused by excessively wet weather.

GALLS. Bacteria causes wartlike growth on the crowns, roots, and canes of plants.

HALF-FORMED FRUITS. The tarnished plant bug is the villain here.

HARD FRUIT. Usually caused by mosaic virus.

LEAF CURL. Caused by an aphid-spread virus disease.

LEAF HOLES (see Cane Spots, above).

NO FRUIT. If the fault isn't improper cultivating, the major culprit is mosaic virus disease, which causes sterility and which can only be avoided by using certified virus-free plants.

ORANGE LEAF RUST. A disease spread by wild raspberries in the area.

WILT. A fungus disease that affects and is transmitted by members of the potato family and can be carried over in soil in which they were planted.

WORMY FRUIT. Caused by raspberry fruitworm.

YELLOWING LEAVES or the YELLOWS. Another mosaic virus disease spread by aphids.

NEW PLANTS FOR NOTHING

RED AND YELLOW varieties of raspberries send up an abundance of suckers connected to the mother plant's roots. These suckers can be severed from the fruiting canes, dug up, and transplanted to another location. This operation might be performed in early spring or late fall, but the best time to do it is in June when the suckers are 6 to 8 inches tall. Larger suckers from the previous year can also be transplanted. In either case, be sure to get a piece of the old root. Do the transplanting during cloudy, moist weather if possible, avoiding hot, dry days, and water in the new plants well.

REDS AND YELLOWS are also propagated by root cuttings. Here 2 to 3-inch pieces of root are cut from established plants in early spring. These are scattered on the surface of a nursery bed and covered with 2 inches of soil. The next spring new plants that come up from the root cuttings can be set out in the garden.

Some *black* and *purple* raspberry varieties produce a few suckers, but these types are usually propagated by what is called "tipping" or tip-layering, which mimics their natural way of increasing themselves. In spring, pinch off the tips of the canes when they are 12 to 18 inches high. This will make the canes branch freely and form a large number of snakelike tips for burying in the soil. Toward the end of August, loosen the soil around the

A root cutting of a red raspberry planted in about 2 inches of good soil will produce a new plant next spring. Dotted lines indicate where cuts should be made.

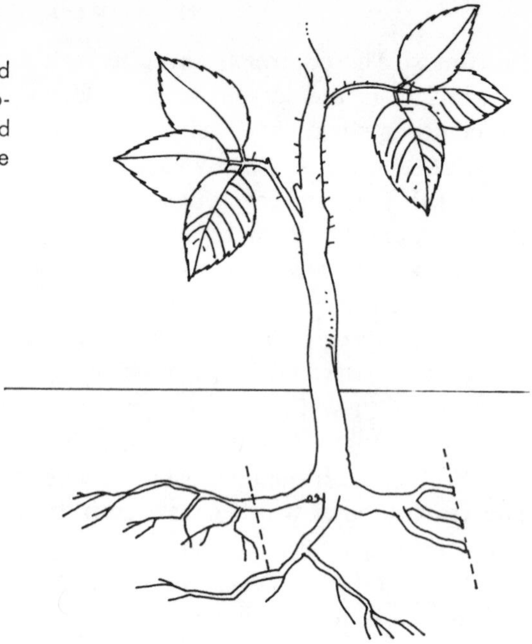

Tip-layering a black raspberry plant with a stone. The tip will take root and send out new shoots.

plants and point the tips straight downward in the soil about 3 inches. Firm the soil around each tip and wait until the next spring, when each tip will be well rooted and can be dug up, severed from the parent plant (even if the tip plant doesn't show growth on top), and transplanted to a new location or grown in the same spot. Another way to tip-layer is to cut two slits in the bark about 6 inches from the tip, preferably on each side of a bud. Lay the cut portion on the ground and hold it in place with a wire peg, stone, or a mound of soil and wait until it roots. In both cases cut what remains of the parent cane to the ground so that anthracnose disease doesn't spread from the old plant.

All raspberries can be increased by mound-layering (see under Mounding, in the Index), and by seed. Seed is not often used, as raspberries grown from it don't come true to form. But the method is easy enough if you want to gamble for a better variety. Choose seeds from only the best-tasting, largest fruits and separate them from the pulp by washing. Let the seeds dry a little, but not completely. Sown at harvest time in a sandy soil, they will germinate by spring, be ready for transplanting the following spring, and yield some fruit the year after.

Harvesting Raspberries With TLC

Of all fruits, raspberries should be picked the most carefully. The berries are ready to pick when they separate easily from the stem, and berries that are perfectly ripe are soft and juicy and have a fine aroma that they only retain for a day or so. Pick the canes at least twice a week (every other day in hot or wet weather) and try to pick in the late afternoon when flavor is highest, or in the early morning when there is no dew on the fruit. Never harvest berries when they are wet unless heavy rains threaten to ruin the crop—and do so with even greater care if you must.

Use the thumb, index finger, and middle finger to pick raspberries, slipping them from the stem—those that don't slip off easily shouldn't be picked, for raspberries do not ripen off the plant. Since they are very tender, gently place them in a shallow container (deep baskets in which raspberries are piled in more than 2 to 3 layers cause unnecessary crushing of the fragile berries). Take care to discard or separate diseased, injured, or overripe fruits. Do not keep the fruits in your hand after picking and be sure to shade the container from the sun.

Raspberries are better left unwashed—washing dilutes their flavor—but if pesticides have been used and they must be cleaned, plunge them in very cold water and remove them quickly.

Raspberry Recipes

Dozens of delicious dishes are made from raspberries—everything from brandies to raspberry pie. Try not to waste fresh raspberries on these recipes, however, as good as they are—eat the fresh ones with a little sugar and/or heavy cream and use frozen raspberries instead whenever possible.

Raspberries are easy to freeze if you have a surplus. Just select firm, ripe berries, wash in ice water quickly, and thoroughly drain them before freezing in pint or quart containers (or don't wash them at all if you use a clean mulch and no sprays have been used on them). To make sure that no berries get mashed together, freeze a layer at a time on a flat tray and then transfer the frozen berries to the container.

The following are recipes that can be made with either fresh or frozen berries unless otherwise indicated.

Quick Sugarless Raspberry Jam

Remove seeds from one cupful of berries by rubbing through a sieve. Add honey to taste and one package of pectin, put in blender, and blend for 2 to 3 minutes briskly. The jam won't keep long but is delicious.

Raspberry Shrub

An old American summer drink, dating from at least the eighteenth century, that is made from raspberries, cider vinegar, and sugar. To four quarts of raspberries add one quart of cider vinegar. Let stand four days, strain, and to each cup of juice add a cup of sugar. Boil 15 minutes and bottle when cold.

Raspberry Sauce

½ cup sugar
½ cup water
1 cup raspberry jam
2 teaspoons cornstarch

Combine sugar and ½ cup water. Bring to a boil, stirring until sugar is dissolved. Add jam and cornstarch, blended to a smooth paste with 2 tablespoons cold water. Bring again to a boil. Remove from heat. Strain. Excellent on ice cream.

Old-fashioned Raspberry Vinegar

"Put a quart red raspberries in a bowl. Pour over them a quart of strong apple vinegar. After standing twenty-four hours strain through a bag and add the liquid to a quart of fresh berries. After twenty-four hours more, strain again, and add the liquid to a third quart of berries. After straining the last time, sweeten liberally with sugar, refine and bottle." An alternate method is to pour red wine vinegar over three inches of red raspberries in the bottom of a sterilized quart bottle with an air-tight stopper. Plug up the bottle and the vinegar will be ready in 5 to 6 days.

Raspberry Pie

Use the same recipe as for blueberry pie (see Index), substituting three cups of fresh raspberries for the blueberries.

Raspberry Parfait

1 pint fresh raspberries
1 six-ounce package sugar wafers, crushed
1 pint lemon sherbet

Crush raspberries, saving a few for garnish. Layer crushed berries with wafers and lemon sherbet in four parfait glasses. Freeze.

Raspberry Mousse with Marshmallows

1 cup sweetened raspberry juice
1 cup sugar
20 marshmallows, diced
Juice ½ lemon
1 cup heavy cream, whipped

To obtain juice from berries, crush a pint of raspberries slightly, add 1 cup sugar, and let come to a boil slowly. Stir to keep from burning. Press through a sieve and measure. Add water to make a cup of juice if necessary. While hot, add marshmallows and dissolve. Cool and add lemon juice, then fold into the whipped cream. Freeze.

Raspberry Delight

1 package frozen raspberries
2 tablespoons lemon juice
½ cup sugar
1 cup thick sour cream

Crush raspberries. Add all remaining ingredients. Pour into freezing trays; freeze for 2 hours, stirring twice during freezing.

Raspberries Supreme

½ cup heavy cream
2½ cups stewed raspberries
½ cup shredded toasted almonds
Pound cake

Whip cream and combine lightly with drained berries and ¼ cup almonds. Line a shallow dish with thin slices of cake. Spread with half the fruit mixture. Add another layer of cake and remaining fruit. Sprinkle with rest of almonds. Chill and serve with cream.

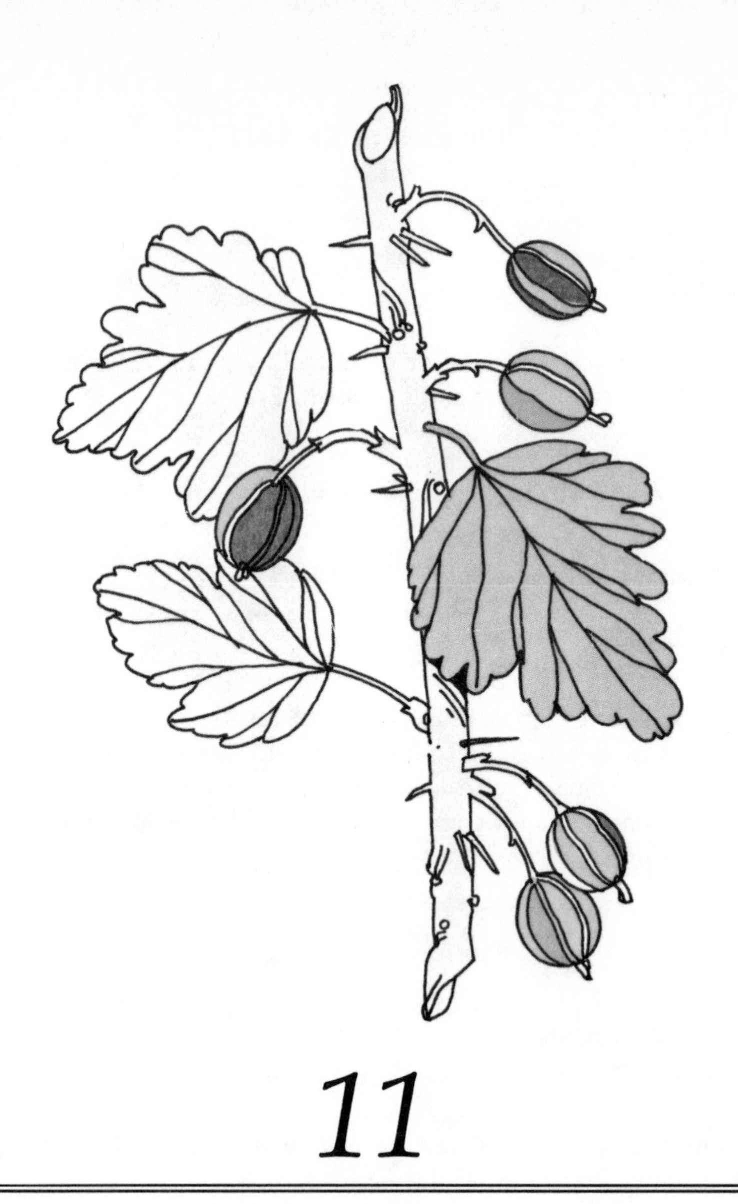

11

A Gander at Gooseberries

Few Americans have ever seen or tasted a gooseberry, let alone grown a gooseberry bush or tree. Rarely offered in even gourmet fruit stores and almost never listed on the menus of even the poshest Continental restaurants in the United States, the delicious fruit deserves far more popularity than it has among Americans. The luscious berries themselves are sometimes as big as eggs, and grow on handsome ornamental shrubs that bear heavily a short time after planting, take up little room, are easy to train as hedges or fence rows, require little care, and, if lack of sun is a problem in your garden, are the most shade-tolerant of all fruits.

Unlike their American counterparts, British gardeners do not neglect the gooseberry; in fact, they have a taste for the fruit that borders on the fanatical—over the centuries they have probably developed close to two thousand varieties of gooseberries with varying flavors and green, pink, red, white, and yellow skins. Gooseberries have been cultivated in England since the early sixteenth century. They had been raised elsewhere in Europe, Asia, and Africa some two thousand years before that and no one knows if the fruit was indigenous to Britain or if it was introduced at an early period and became naturalized, but the British certainly have done far more to improve the species than gardeners anywhere else.

How the gooseberry got its name is a puzzle to historians and etymologists. One theory claims that "gooseberry" is a mispronunciation of "gorge berry," an early name for the fruit; a nice theory, except that "gooseberry" was used before "gorge berry," according to what records there are. For similar reasons, most word detectives do not believe that the "gooseberry" is a corruption of *groseille,* the French name for the fruit, or the Dutch *kruishes,* which means "cross berry." Perhaps a better explanation is that "gooseberry" is a corruption of the German *Jansbeeren* (John's berry, so named because it ripens during the feast of St. John), which corrupted into the German *Gansbeeren* and was translated into English as "gooseberry" because *gans* means "goose" in German. That, in fact, is about the only thing linking geese with the berry or plant—the goose doesn't like gooseberries, isn't even averse to them, just ignores them entirely. Neither were gooseberries customarily served with roast goose, as is often stated. The consensus is that "gooseberry" does come from some unknown association with the goose, plant names associated with animals commonly being inexplicable. There is even a theory which holds, put simply, that the goose gave its name to a fool or simpleton (as in "a silly goose") and that the green of the berry (suggesting a "greenhorn" or fool) became known as a goose (or "fool") berry.

No matter how its name evolved, the British have celebrated the gooseberry in song and story as well as in actual fact. Gooseberries were so common in Elizabethan England that Shakespeare used the expression "not worth a gooseberry." There were in early times "gooseberry shows," "gooseberry fairs," and "gooseberry feasts," and the fruit was used in scores of dishes. It was an old Norman practice to eat green gooseberry sauce with mackerel, and for this reason gooseberries were called *groseille à maquereau* in French to distinguish them from currants, both of the fruits being *groseilles*. Pigeons and other fowl were stuffed with gooseberries, which were and still are prized for eating fresh out of hand when dead ripe, as well as in gooseberry pies, tarts, pastries, puddings, jellies, jams, and even a wine called "gooseberry" celebrated in poems by both Oliver Goldsmith and Charles Lamb.

Probably the most famous dish made from gooseberries is "gooseberry fool," a dessert made of the fruit stewed or scalded, crushed, and mixed with milk, cream, or custard. Some say the "fool" in the dish is a corruption of the French verb *fouler,* to crush, but this derivation seems to be inconsistent with the use of the word. More probably the dish is simply named after other, older fruit trifles, the use of "fool" in its name in the sense of "foolish or silly" being suggested by "trifle." In any case, gooseberry fool has been an English favorite since at least 1700 and Mrs. Glasse gives a recipe for it in her famous *The Art of Cooking* (1747). So widely known is the dish that another *plant* is named after it, the English calling the willow herb *Epilobrium hirsutum* "gooseberry fool" because its leaves smell like the dessert!

One British writer thought that when fully ripe gooseberries tasted nearest to grapes of any other fruit, but that description leaves much to be desired. The truth is that gooseberries have a unique flavor of their own beyond compare. They have been paid compliments by many discerning writers, but the words of little Marjorie Fleming, "Pet Marjorie," the youthful prodigy of Sir Walter Scott, are most memorable. Wrote Marjorie in her quaint and charming diary shortly before her death at age seven: "I am going to turn over a new life and am going to be a very good girl and be obedient . . . here there is plenty of gooseberries which makes my teeth watter."

English gooseberries were brought to America by early colonists and cultivated varieties were featured as minor offerings in our earliest garden catalogs, but the English fruit never caught on in America. Neither did our smaller-fruited native species. There were several reasons for this: thorny bushes; the need of the tastier English varieties for a milder climate with less extremes of cold and heat; small and inferior fruit on the native types; and the fact that gooseberry bushes host a serious disease of white pine trees. But certainly the reason the fruit didn't succeed wasn't that Americans dislike the taste of gooseberries. Green gooseberry pie, gooseberry relish, spiced gooseberries, and plump, juicy gooseberries eaten fresh off the bush were early American favorites that deserve a revival. With the new, improved varieties offered today, all gardeners, even those in hot climates, can grow the delicious fruit, so there is no reason not to try them.

WARNING!

Before beginning a discussion of varieties and cultural methods, the gardener should know that gooseberry bushes serve as hosts to a stage of white pine blister rust ("Pine Forest Disease"), a fungus accidentally imported into America at the turn of the century that kills valuable white pine trees. This fungus, which lives on gooseberry bushes and currants for part of its life cycle, at one time caused at least twenty-four states to forbid the planting of gooseberries or require a permit before anyone could grow the fruit. Today only Idaho, New Hampshire, New Jersey, Vermont, and Washington forbid the shipping of gooseberries into their jurisdictions without a permit, and dealers cannot sell you gooseberry bushes if you live in one of these states. Unless you decide to transplant gooseberries from the wild, nurseries will solve this problem for you, since they can't ship gooseberries to areas with the prohibition. Sometimes, however, only certain *areas* in states are restricted, and gooseberries not infected by blister rust fungus are allowed to be planted elsewhere. In any event, restrictions or no, *don't plant gooseberries within 1,000 feet of white pine trees* (pines with five needles in a bundle) *or within 1,500 feet of any area where white pine seedlings are being grown.* You're only asking for almost inevitable trouble if you do.

The Best American Gooseberries To Grow

All gooseberries are prolific, long-lived plants that bear when they are 2 to 3 years old and live up to 20 years—some plants growing under ideal conditions have lived over 40 years, and two famous English specimens growing against a garden wall reached the age of 60 years, each measuring over 50 feet from one extremity of the branches to the other. Three bushes should provide plenty of fruit for a family of four, each plant producing 3 to 4 quarts of berries at maturity. All varieties are self-fruitful; that is, one gooseberry bush doesn't require other gooseberry bushes planted nearby to ensure pollination and fruit. There are two quite distinct species that can be classified as American and English, and while the English are certainly by far the superior in taste and size, many American varieties have been developed that are well worth growing.

RIBES HIRTELLUM, as the American gooseberry is called, includes numerous named varieties, some produced by hybridizing the wild American species with English varieties and others simply hybrids of our native wild gooseberry. Following are some of the better ones, listed in the order that most experts rank them. Generally, the reds are considered to be the sweetest.

PIXWELL. Prolific, pinkish red upon ripening, and the large fruits are "underslung" on the branches, hanging away from the thorns and making for easy picking. Developed as a hybrid of the American wild gooseberry *R. missouriense* by the North Dakota Experiment Station, Pixwell is a compact bush that will thrive anywhere gooseberries can be grown.

WELCOME. This University of Minnesota hybrid has very few thorns. Light pink when fully ripe, it is often picked when green for use in pies and preserves.

JOSSELYN (see Red Jacket, below).

POORMAN. A productive plant with large-fruited, pear-shaped, bright wine-red berries that are very sweet when dead ripe, though the skin is rather tough. The upright to spreading bushes are larger than most varieties and noted for their freedom from diseases. Many experts say this is the best-flavored American variety of any color.

DOWNING. The best green-fruited variety, with sweet flesh, tough skin, and medium size. A vigorous, prolific, upright plant, Downing is one of the few gooseberries still offered in America that was grown here before 1900.

RED JACKET. Earlier than most American gooseberries, but smaller in size than most, too, this sweet, red-fruited variety is sometimes offered as Josselyn.

Other American named varieties worth a try are Oregon, Como, Pearl,

Houghton (another very old type), and Glendale (another hybrid of the native wild gooseberry).

American gooseberries are offered by many nurseries, including Burgess, Farmer, Field, Foster, Gurney, Miller, New York Co-op, Shumway, and Southmeadow, their addresses given in Appendix III.

The Best Foreign Varieties

English gooseberries (*Ribes grossularia*) are more difficult to grow in America than the native hybrid types, but since they are perfection in gooseberries, many advanced gardeners will want to try them. They generally do not thrive except in the cool, dry regions of the Pacific coast, though gardeners on the East Coast and in other areas have raised them successfully with a little care. Though they aren't the gooseberries to begin gooseberry gardening with, a little time growing the fruit and studying its habits, combined with the hints given here, should enable most Americans to raise the fruit successfully.

The latest statistics I can find indicate that the English devote some 13,000 acres to gooseberry growing while Americans only plant 900 acres in the fruit. English gooseberry fairs over the past three centuries encouraged the development of thousands of varieties there. These fairs, offering prizes for the best fruits, led to the introduction of gooseberries with red, pink, yellow, green, and even white skins. Some were developed for flavor, others for size; some for eating out of hand, others for cooking. Generally speaking, the yellow-skinned types are most highly valued for flavor when eaten out of hand.

The incredible number of English gooseberry varieties makes a complete listing here impractical. Chautaugua (yellowish-green), Fredonia (red), Industry (red), and Columbus (red) are the English varieties most frequently grown in America. Varieties that the British prize are the large Careless, a white-skinned dessert type; Keepsake and May Duke, early types; and the dessert types Leveller, Shiner, Lord Derby, Gunner, Leader, White Lion, and Conson's Seedling. Other famous English varieties are Achilles, a very large, late-ripening type; Catherina, a large, sweet, egg-shaped, golden-orange berry; Early Sulphur, said to have "a fine apricot-like aftertaste"; the large Whinham's Industry, which reportedly "does well under trees"; Whitesmith, one of the best greens; and the Worcesterberry, a large, blackish berry that may be a cross between the gooseberry and black currant. English gardeners have grown gooseberries up to a record weight of over two ounces and the size of a small apple—about 6½ inches in circumference.

In addition, the Canadian Experiment Station has originated some thornless types from English plants, including the relatively thornless Canada 0-273 and Sylvia and the completely thornless Captivator—all reds and delicious. Excellent, hardy, Finnish varieties include the very large, sweet Hinnomaki Yellow, which survives the severest of Finland winters and the big-bushed Lepaa Red, which is resistant to mildew.

A tree form of English gooseberry has been produced in America by grafting *Ribes grossularia* on vigorous understock of the mountain currant, *Ribes aureum.* The best source for English, Canadian, and Finnish gooseberries in the United States is the Southmeadow Fruit Gardens (see Appendix III), which offers a full 16 varieties. Another source would be English nurseries like Thompson and Morgan, which may be able to put you in touch with English growers who specialize in gooseberries. These specialists might also be able to supply you with bushes of the following highly recommended types, which I've found in old English gardening books but can locate nowhere. The asterisked varieties have been recommended by writers for their large size, while all others have been praised for their flavor.

Reds

Conquering Hero,* Crown Bob,* Dan's Mistake,* Dr. Hogg, Henson's Seedling, Ironmonger, Keen's Seedling, Lion's Provider, London,* Miss Bold, Monarch, Plough Boy, Raspberry, Red Champagne, Red Turkey, Red Warrington, Rifleman, Rough Red, Wilmout's Early Red, Wonderful.*

Yellows

Broom Girl, Criterion, Drill,* Fanny, Garibaldi,* Gipsy Queen, High Sheriff, Lord Rancliffe, Moreton Hero, Mount Pleasant,* Peru,* Rumbullion, Smiling Beauty, Yellow Bill, Yellow Champagne.

Greens

Glenton Green, Green Gascoigne, Green London,* Green Overall, Green River, Green Walnut, Gregory's Perfection, Heart of Oak, Hebburn Perfection, Jolly Anglers, Keepsake, Laurel, Lord Eldon, Pitmaston Greengage, Random Green, Roseberry, Stockwell, Telegraph,* Thumper,* Thunder.

Whites

Abraham Newland, Adam's Snowball, Antagonist,* Bright Venus,

Careless,* Cheshire Lass, Crystal, Early White, Hero of the Nile,* King of Trumps, Lady Leicester, Mayor of Oldham, Princess Royal, Queen of Trumps,* Royal White, Snowdrop, White Champagne, White Fig, Woodward's Whitesmith.

Gooseberries To Gather

The name "gooseberry" has been applied to many fruits growing in the wild. A number of these, like the Indian gooseberry (*Vaccinium frondosum*), are not true gooseberries despite their resemblance to the plant and have no place here. But others deserve mention as possible garden curiosities, or for those who wish to gather their gooseberries in the wild. I know of no nurseries offering these, so only gatherers willing to search the woods can sample them or bring back cuttings to plant:

RIBES HIRTELLUM. The *American wild gooseberry* or *currant gooseberry*. Small, purplish-black fruits less than half an inch in diameter on a very thorny plant.

RIBES OXYCANTHOIDES. The *wild mountain gooseberry*. Has reddish-purple berries half an inch in diameter.

RIBES ROTUNDIFOLIUM. Good-flavored, glabrous, purple fruit on a plant armed with small, inconspicuous, solitary thorns.

RIBES CYNOSBATI. The *prickly gooseberry*. Dark-brown, eatable berries on a thorny plant found in the Catskills and North Carolina.

RIBES DIVARICATUM. Small, purplish-black, glabrous berries one third of an inch in diameter on a large shrub that grows up to 10 feet tall and is armed with large thorns nearly an inch long.

RIBES VALDIVIANUM. Very similar to the above except that the shrub grows to only about 6 feet tall and is more branching in habit.

RIBES LACUSTRE. The *swamp gooseberry*. Thoreau mentions this native plant in his *Maine Woods*, recalling how he "saw the swamp gooseberry with green fruit."

When, Where, And How To Plant

Autumn or early spring is the best time to transplant gooseberries, the earlier the better in spring, as the buds open with the first seasonal warmth. Gooseberries tolerate cold well and prefer a moist, partially shaded spot in the garden. Shade is particularly important in hot climates, where the plant finds it tough going. Remember that gooseberries can stand more shade than any other cultivated fruit; about a half day's sun is all they need. The English varieties especially dislike hot American summers and usually refuse to grow well except close to the Great Lakes and along the northern Atlantic and Pacific coasts. You might have good luck with them elsewhere, however, by planting bushes on the north or east sides of buildings, fences, hedges, or arbors, or even by planting them in the shade of trees.

Gooseberries flourish in many soils, but rich, moist, well-drained clay loams yield the best fruits. The plants generally do poorly on light loams or in sandy soils, and in dry soils they often suffer from premature falling of fruit. Though gooseberries will survive in the average good garden soil that grows flowers and vegetables, try to select a rich, moist soil. If this isn't possible, remember the old gardening maxim that it is better to plant a ten-cent plant in a dollar hole than a dollar plant in a ten-cent hole. Work plenty of well-rotted manure and compost into the planting area and keep it well watered and mulched all summer long.

One gooseberry bush for every person in the family is recommended, since a self-fruitful gooseberry bush yields 3 to 4 quarts of berries. Order healthy one-year-old bushes for planting and set them into the ground 4 to 6 feet apart, in rows 5 to 8 feet apart if more than one row is planted. The gooseberry is a compact bush that usually doesn't need the greater distances between plants except for very vigorous varieties, which are generally the midseason types. Five feet apart each way is a good average distance, if your supplier doesn't specify otherwise. All varieties should be pruned of damaged roots and set in their planting holes with their lower branches a little below the soil to encourage them to grow into a bush form. Once they are planted, water deeply and mulch the bushes to a height of 3 inches with sawdust, hay, straw, well-rotted manure, corncobs, or grass clippings to preserve moisture and keep the roots cool. (See Appendix II for other mulches.)

Space Savers

Gooseberries do well in partial shade, and there's usually shady space to spare in most gardens. Since the bushes grow only about 3 feet high and take up very little space, a single gooseberry, with its attractive foliage, can be set in among foundations or border plantings. Or a bush can be planted in a fence row, or between young fruit trees and grapevines.

Another space-saving alternative is to prune gooseberry plants to a single stem and grow them as standards, which do not yield as well but give bigger berries and can be planted much closer together in the garden row. The bushes can also be trained as espaliers: against a wall with a northern aspect where summers are very hot, or on a southern wall where there isn't much sunshine.

PRUNING

Gooseberries bear fruit buds at the base of one-year-old wood and from spurs on older wood. The bushes will yield for a long time without pruning, but will bear more fruit and larger berries if pruned every year. Pruning should be done when the plants are dormant—after the berries fall in autumn and before growth begins in spring. The first year, after the fruit is picked, cut out all weak shoots at the base, leaving about six strong branches. In subsequent years remove all branches older than three years and limit each plant to about fifteen strong branches, cutting out any dead or weak wood and branches very close to the ground. Since the best fruit is borne on two-year-old canes and shoots from three-year-old canes, this pruning practice will ensure good quality. After the third year, a well-pruned bush should ideally have five three-year-old canes, five two-year-old canes, and five one-year-old canes of the previous summer's growth.

Pruning is sometimes necessary to prevent powdery mildew on gooseberries. European varieties are especially susceptible to this disease, which can often be discouraged by keeping an open head on the bushes; that is, by cutting out twigs to develop open-spreading tops that aren't too dense. For the same reason, cut out all branches that lie too close to the ground. On the

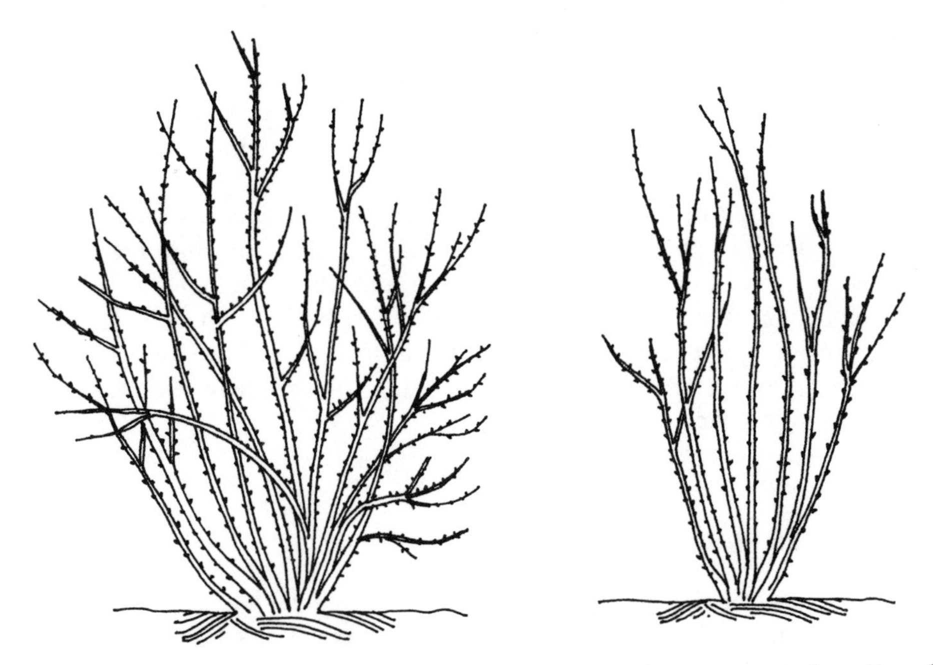

Prune gooseberries every year, removing three-year-old wood to encourage formation of fruit-bearing wood. At left is an overgrown bush, at right a well-pruned one.

other hand, gooseberries grown in the South should *not* be pruned too thin, as a dense-headed bush will provide shade for the fruit.

When gooseberries are trained as espaliers against a wall, the fan system of training may be adopted, part of the old wood being pruned annually. Or young plants can be planted 3 feet apart along the wall and pruned so that three shoots from each run perpendicularly along the wall at equal distances.

Where gooseberries are not thinned by pruning—and some British gardeners prefer to do no pruning at all—branches heavily laden with fruit on pendulous varieties can be prevented from touching the ground by propping them with forked sticks.

Care Of Gooseberries

Fertilizing

Though gooseberries are heavy feeders, they will need no additional fertilizing the first year if plenty of decayed manure is worked into the soil at planting time. Every autumn thereafter work manure, bone meal, or cottonseed meal into the soil near the outer branch line of each plant, being careful not to disturb the shallow roots, and replenish the mulch around them. Other choices are to fertilize with 8 ounces of nitrate of soda per plant, or 4 ounces of ammonium nitrate per plant, or a couple of handfuls of 5-10-5 fertilizer per bush. Where soils are very light, an application of sulfate of potash at about 4 ounces per plant every other year will prove helpful.

A good way to promote deep root growth in gooseberries is to fertilize down to a depth of about 8 inches by forcing a spading fork into the ground in a circle around the outer edge of a plant and filling the holes with fertilizer. Remember to use only *one* of the above fertilizing methods every year. Overfertilizing gooseberries often results in lots of foliage and no fruit. Should this prove a problem, try not fertilizing for two years and see if the bushes begin to bear.

Watering and Weeding

Use a deep mulch and practically all watering and weeding of gooseberries is eliminated. Manure, sawdust, or hay make the best mulches for the plants (see also Appendix II). Be certain, however, that the mulch is restored from year to year, as it breaks down, and water occasionally so that the ground never dries out around the bushes. If you choose not to mulch plants, water frequently and weed the plants carefully, never disturbing the roots by cultivating too deeply.

Insect Pests and Diseases

You may never be bothered with insect pests or diseases on young gooseberry bushes, but the plants are not immune to predators. Here are the most troublesome ones and ways to combat them.

POWDERY MILDEW. Furry-white to light-brown patches on the leaves, canes, or fruit is usually evidence of this fungus disease, which occurs in wet areas, generally in spring and fall, and can cause defoliation and thus a sparser crop the next year. As noted, pruning will discourage powdery mildew—plants with at least a half day of sun and adequate air circulation aren't severely affected by it. The disease can also be controlled by spraying with either Bordeaux mixture; dormant spray of lime sulfur; the fungicides Karathane or Miragard; or various all-purpose fruit-tree sprays. Be sure to follow the manufacturer's directions carefully when using any of these sprays. And burn all branches infected with powdery mildew that are pruned from gooseberry bushes.

LEAF SPOT DISEASE. Dark spots on older leaves, canes, and fruits, beginning on lower leaves and resulting in the defoliation of plants toward the end of summer if not treated. Often occurs in wet areas. Spray the plants with Bordeaux mixture as soon as leaf spot disease appears. The following year spray three times: after blossoms fall, after harvesting fruit, and three weeks after harvest. Help prevent the disease by refraining from fertilizing too heavily with nitrogen.

ANTHRACNOSE. Use the same treatments as for Leaf Spot Disease, above. *Welcome* is a particularly anthracnose-resistant variety.

CURRANT BORER. Yellowish-white grubs or caterpillars that bore inside the canes in spring, leaving dead, hollow canes with black centers and causing the dwarfing of the plant. The only control here is to cut off the infected canes at ground level and burn them immediately.

IMPORTED CURRANT WORM. Large (up to one inch long) green worms with black spots that attack in early summer and can strip a bush of leaves in a few days. Handpicking the worms is possible if you have only a few bushes, but extensive plantings are best sprayed with Sevin or dusted with Garden Guard, rotenone, or pyrethrum as soon as damage is noticed.

CURRANT FRUITWORMS. Small (⅓ inch), tapering, whitish maggots of small yellow flies. Adults are gray moths. The worms bore into fruit, causing them to redden prematurely and drop off. Controls are the same as for the Imported Currant Worm, above.

CURRANT APHID. Small, yellowish to dark-green plant lice, which suck juice from the undersides of leaves, causing them to become crinkled and red and curl downward. Probably the worst pest of gooseberries, aphids can sometimes be controlled by handpicking the curled leaves and destroying them, or by spraying the leaves thoroughly (on both sides) with an organic spray made of garlic, oil, and green soap. Ladybugs introduced into the garden will also control them. Chemical controls include a dormant lime sulfur spray used before the buds swell in spring to kill the aphid eggs, and malathion 50 per cent spray used when the insects appear.

SCALE. Various types (brown, elongate, tapering, or round waxy scale) under which the tiny insects live—usually at the bases of canes. Spray with superior-type dormant oil emulsion before growth starts in spring, and repeat, using low dosage, just before and after blooming. Ladybugs introduced into the garden also help control scale, feasting on the insects.

MITES. Use the same treatment as for Scale, above.

BIRDS. Birds sometimes damage gooseberry fruits but can be thwarted by netting the fruits as they ripen. Birds do far more good than bad, however, devouring currant worms and other pests. (See Appendix I, For the Birds.)

PROPAGATING GOOSEBERRIES AND CREATING NEW VARIETIES

Gooseberries can be increased by seeds, hard- and softwood cuttings, or layers. Propagating the fruit by seed is only done to originate new varieties, as none of the existing varieties reproduce themselves true from seed; often, in fact, the fruit of seed-propagated gooseberry bushes is a different *color* than that of the parent plant. If you want to create new gooseberry varieties, just take seed from a ripe fruit, wash it thoroughly, removing the pulp, and dry it on a piece of paper. Sow the seed in the open ground in spring after all danger of frost, covering with about an inch of light soil. The young seedlings will be big enough for transplanting the following autumn and should fruit after about three years of growth.

Propagation by cuttings is the commonest method of increasing gooseberries, though the cuttings do not always "take." Hardwood cuttings are generally used, these taken in early autumn, but gooseberries will also root from half-ripe cuttings in summer. In the fall, strong, well-ripened, one-year shoots should be selected and cut off at their junction with the older wood. Trim each cutting to about 10 to 12 inches long and remove all eyes or buds from the lower half of the cutting. Insert the cuttings about 6 inches apart in a 4-inch-deep trench in open, moist ground, and then fill in the trench. Another method is to store the cuttings in sand, sawdust, or peat moss in a cool place over the winter and then plant them in the spring. In either case, give them 1 to 2 years in a protected location before transplanting to their permanent place in the garden.

Layering is the easiest way to propagate gooseberries. One method is simply to lay a large branch on the ground, peg it into the earth with a clothespin (or wedge it down with a rock), and cover it with light soil, taking care not to bury the tip. The new plant should be rooted and can be moved to another location the same season. An alternate method is to mound-layer one old plant. In about mid-July slit the stem bark on as many branches as you want new plants from. Then mound soil over the base of the entire plant, covering the wounded areas on each branch. When roots form in the wounded places, usually by the following autumn, the newly formed plants are severed from the mother plant and set out in their permanent location. The major problem with all types of layering—something adherents of the method seldom point out—is that plants from layers are not so symmetrical as those raised from cuttings.

Harvesting Gooseberries

Except for the thornless varieties and "underslung" types like Pixwell, gooseberries are the most hazardous fruit to pick, since the canes and shoots are adorned with abundant thorns. Each gooseberry bud produces 1 to 3 berries, not big clusters of berries like currants. The bushes bear at 2 to 3 years old and are producing big crops by the time they're 4 to 5. The best way to harvest them is to don a pair of heavy work gloves, steadying a cane with one hand and stripping the berries off with the other—the few leaves that come off with the berries can be removed from the basket later. Some gardeners combine pruning with picking, cutting off branches that must be pruned and taking the severed branches to a table where they can sit down and pick at ease. Both the stems on one end of the gooseberries and the blossom collars on the other end should be removed when the berries are prepared for eating or recipes.

Gooseberry Delights

Here are some traditional gooseberry recipes—many of them British favorites for centuries—and a few mouth-watering innovations as well.

Gooseberry Fool

1 pound green gooseberries
½ cup sugar
1 pint water
1 cup heavy cream, whipped

Stew the gooseberries with sugar in one pint of water until they are very soft. Purée through a food mill, taste for sweetness, and chill before serving with whipped cream.

Gooseberry Tarts

Wash 4 cups of gooseberries and drain. Cook slowly with 1½ cups sugar and ¼ cup water until berries are tender, stirring constantly until sugar is dissolved. Add 1 tablespoon solid shortening and 1 teaspoon grated lemon rind. Pour into baked tart shells and garnish with whipped cream.

Gooseberry Jelly

Boil 6 pounds of unripe green gooseberries in 6 pints of water until soft. Pour them into a bowl and let stand covered with a cloth for 24 hours. Then strain through a jelly bag, and to every 2 cups of juice add 1 pound of sugar. Boil the jelly 1 hour, then skim it and boil for ½ hour longer.

Gooseberry Soy

Take 6 pounds of gooseberries that are nearly ripe and 3 pounds of sugar, 2 cups of vinegar, and boil all together until quite thick. Season to suit your taste with ground cloves and cinnamon. This is eaten with meats and keeps a long time.

Gooseberry Catsup

Eight pounds of ripe fruit, 2 cups of vinegar, 4 pounds of brown sugar, 2 ounces each fine cloves and cinnamon tied in a bag. Cook the berries, vinegar, and sugar over medium heat for 3 or 4 hours, then add the spice, boiling a little more. Put the catsup in a jar and cover it well.

Pickled Gooseberries

Add 1 pint of vinegar and 3 pounds of sugar to 6 pounds of gooseberries. Cook over medium heat 20 minutes. Add 3 more pounds of sugar and cook 20 minutes longer. Seal in glass jars.

12

The Electrifying Currant

Currants are another rare, shade-tolerant fruit that form the basis for many mouth-watering dishes few Americans are familiar with. These smaller relatives of gooseberries, which are cultivated in much the same way and add a brilliant touch of red to the garden, weren't always as unknown in America as they are today. English settlers brought the tart red currant to the Massachusetts Bay Colony in 1639 and housewives were soon using it to make jams, jellies that were served with venison, pies, pastries, and boiled puddings. Wrote poet Richard Hughes in the early 1600s:

Puddings should be
Full of currants, for me:
Boiled in a pail,
Tied in the tail
Of an old bleached shirt:
So hot that they hurt.

Europeans were enjoying currants long before Americans, of course, but the fruit was a wild one until the 1500s, when it began to be cultivated for the first time in the Netherlands, Denmark, and around the Baltic Sea. The currant got its name in a rather roundabout way, having been named, in fact, after a variety of raisin. It seems that in the fourteenth century Corinth, Greece, was the chief port for exporting seedless raisins dried from a small variety of grape grown in eastern Mediterranean lands. These raisins were originally called raisins of Corauntz—Corauntz being the Anglo-French pronunciation for Corinth—and finally *Corauntz* was corrupted to *currants.* The black, red, and white berries of the Ribes family that are known today as currants were given the same name as the raisins because the plant clusters of the black variety of *Ribes* look something like the small grapes that were made into raisins of Corauntz.

Strangely enough, the currant's family name, *Ribes,* also stems from a confusion of identities. *Ribes* derives from the Arabic *ribas,* a rhubarb grown by the Arabs in Lebanon as a medicinal essence. When in the early eighth century, the Arabs conquered Spain and found themselves without their trusty *ribas,* they looked for something to replace rhubarb and selected currants and gooseberries, which they also called *ribas.*

Black currants (*Ribes nigrum*) have never been as esteemed by American gourmets as red and white currants; in fact, centuries ago they were thought to breed worms in the human stomach. But they are high in vitamin C and have had many therapeutic properties attributed to them over the years, having been recommended for arthritis, gout, and dropsy, among other complaints. The black fruit has sweet, reddish flesh; a famous concoction made from it is the French cassis de Dijon, an internationally famous liqueur.

Red and white currants are also high in vitamin C and have been considered refreshing medicines as well as a cure for dysentery. Tart, but sometimes eaten out of hand, the berries are mostly used in making jelly, preserves, syrup, and currant wine. The English used to make a dish called "laid pudding" in pre-Reformation days that consisted of layers of raisins and red currants alternating with delicious, custard-soaked teacakes. Yet the most famous dish made from red and white currants has to be the seedless currant jelly once laboriously made by professional "seeders" employing a goose feather to delicately pick out the currant seeds one by one without damaging the berries. The jelly used to be a gourmet delicacy made only in Bar-le-Duc, France, on the banks of the Meuse, and ever since 1559, when Mary Stuart, later Mary Queen of Scots, was given a jar of the rare jelly, it was presented to every visiting chief of state. But just five years ago its last manufacturer went out of business as a result of prohibitive labor costs, and the gastronomical rarity is no more.

Though currants, like gooseberries, aren't exactly the rage in America, they are being planted more frequently presently than at any time since colonial days. Very easy to grow, they take up little room, make incomparable jellies, among other good dishes, and are well worth a try in every home garden. *Remember, however, that currants—especially the black currant—can harbor the fungus that causes white pine blister and cannot be planted in certain areas* (see Gooseberries, WARNING, in the Index).

THE BEST CURRANTS TO GROW

Currants come in black, red, and white (ranging from yellow to white) colors. *No blacks are recommended here* because black currants are carried by few, if any, American nurseries (McFadden Seed Co., in Canada, does offer a nameless black variety). Black varieties that might be ordered from nurseries abroad include Black Naples, Lee's Prolific, Sweet-Fruited, Ogden's Black Grape, Baldwin, Boskoop, Seabrook, Laxton, and Daniels September. Canadian blacks include Climax, Kerry Elipse, Clipper, and Saunders. Black currants are used mostly for cooking or preserves and for their medicinal properties.

Red Currants

RED LAKE. Probably the most popular red variety, at least the one offered by most nurseries. It is a very hardy, vigorous type that bears the first year after planting and yields large crops. Its clusters are long, the berries uniformly large, and the plants are compact and less vigorous than most. Like all reds, Red Lake is generally made into preserves.

WELDER. A spreading, productive plant that bears long clusters of large berries that are pleasantly subacid; bears even more than Red Lake on less fertile soils and the fruit hangs a long time after ripening without going bad.

IMPROVED PERFECTION. A productive improvement of an old variety that has larger berries than either of the above.

Other old favorite reds are London Market, Fay, Diploma, Cherry, Red Cross, Fay's Prolific, Victoria, and Cascade. Old European reds include Rivers Lake, Houghton Seedling, Knights Large Red, Prince Albert, La Fertile, Laxton's Perfection, La Hative, Mammoth, Baby Castle, Warner's Grape, Red Champagne, and Red Dutch.

White Currants

WHITE GRAPE. Traditionally considered the best American white currant. Like most white currants it is mostly used for desserts, being less acid than reds or blacks.

WHITE IMPERIAL. A recent challenger of White Grape recommended by the New York State Testing Cooperative.

HOLLAND WHITE. The long-time favorite European white currant (sometimes called White Dutch).

GATHERING CURRANTS

Currants have been well-known throughout America, as familiar names in American history like the *buffalo currant*, the *skunk currant*, and the *squaw currant* clearly show. There are well over 140 species of currant, but the following are the three species most commonly found growing wild in America. All bloom early in the spring and fruit in midsummer.

RIBES AUREUM. The showy *garden currant,* which has yellow flowers and purplish-brown fruit, growing 4 to 6 feet. Western North America.

RIBES ODORATUM. The *buffalo currant,* a 4 to 6-foot-high ornamental with showy, yellow flowers in drooping clusters and black fruit. Central United States.

RIBES SANGUINEUM. The *flowering currant,* an ornamental bush 8 to 10 feet high, which has red flowers and bluish-black fruit. Northwestern North America.

WHEN, WHERE, AND HOW TO PLANT

Currants do best in a moist, cool climate, but they will yield a crop of some sort almost anywhere in America except the very hottest and driest regions. The fruit can be grown successfully in any good garden soil that grows flowers and vegetables, but the ideal soil for currant growing is a cool, moist, well-drained clay or loam, 2 feet or more in depth and on the heavy side to help it retain moisture. Currants that are planted in light soil and aren't kept watered constantly will produce fruit that shrivels or ripens prematurely and is inferior in flavor. The bushes should never be planted in a southern exposure. Though they are not particular about it, a pH of from 5.5 to 7.0 is best for them.

Plant currants in the late fall or early spring before growth starts, enriching the soil with manure or compost prior to planting. Use one- or two-year-olds to obtain berries within two years and set them about 5 feet apart each way (give black currants 6 feet each way), cutting back the plants to about 8 inches from the ground after planting. Bearing-age transplants will yield berries sooner, but are more expensive and less likely to survive.

One or two currant plants for each member of the family should suffice for all but great jelly-makers. Each bush will yield 4 to 6 quarts annually and live for up to 30 years or more.

Space Savers

Currant bushes only grow 2 to 4 feet tall and spread out about the same distance. Everything written about gooseberry space savers applies to currants as well (see Index). Currants are just as shade-resistant as gooseberries, too, and can be planted in shady areas on your property where no other fruit will grow. A good spot to grow them is between rows of fruit trees or rows of grapes. They can also be planted on the north side of buildings or fences.

To train currants to cover a wall, place young plants three feet apart. Select the three strongest shoots on each plant and train one shoot upright in the center and the other two equal distance on either side at one foot apart. Cut them back if they are at all weak, allowing them to reach the desired height, and occasionally shorten all the laterals.

Care Of Currants

Fertilizing, Watering, and Weeding

Knowing that currants are relatively heavy feeders, some gardeners tend to overfertilize them, and as a result they produce a lot of green growth but few, if any, berries. Currants need not be fertilized the first year if planted properly. Every year thereafter they should be fed in the fall or early spring with a forkful or two of well-composted manure worked very shallowly into the soil around the base of each plant so as not to disturb the roots. If a composted manure mulch is kept around the base of each plant and renewed from year to year, no further fertilization is needed. Currants seem to thrive on organic fertilizers like manure; there is some evidence, in fact, that they do not respond as well to chemical fertilizers. Other organic fertilizers worth trying are composted leaves and cover crops of barley and oats turned under in the fall. If nonorganic fertilizers are used, try either a large handful of a well-balanced, commercial fertilizer scattered beneath each plant in the spring or fall; or 4 ounces of ammonium nitrate per plant; or 8 ounces of nitrate of soda per plant. On light soils, which currants don't like, 3 to 4 ounces of sulfate of potash might benefit each plant if applied every two years or so.

Make sure that the soil does not dry out around currant bushes; that is, water when necessary. Selecting the proper moist soil at planting time and keeping the bushes under a 2-inch mulch of manure, hay, straw, corncobs, or leaves should eliminate any watering problems and hold down weeds as well. If you don't mulch, be sure to cultivate the bushes frequently enough to eliminate weeds, but don't cultivate deeply and harm the currant's shallow root system.

Pruning

Like gooseberries, currants develop from fruit buds at the base of one-year-old wood and from spurs on older canes. Currants, however, yield a cluster of berries (not one or two) from each bud. On planting, currant bushes are best cut back about 8 inches from the ground. From then on,

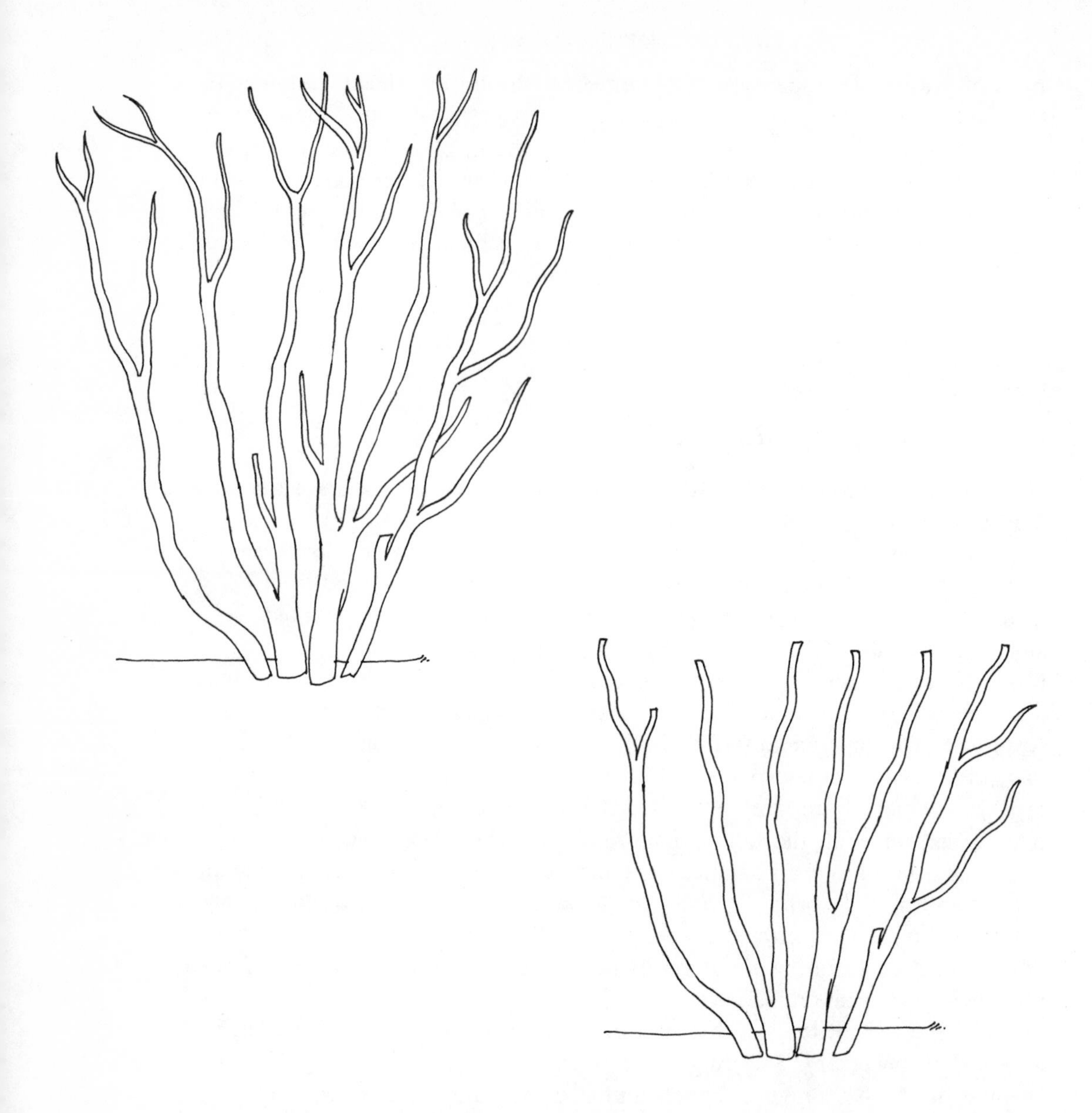

A well-pruned currant bush, (left) the bush before pruning and (right) the bush after proper pruning.

remove any canes that droop to the ground when laden with fruit (these help spread mosaic disease), any cane that doesn't grow at least 6 inches a year, diseased or broken canes, weak one-year shoots, and all canes over four years old (which are usually weak and unproductive). Autumn is the most practical time to prune currants, but the job can be done whenever the bushes are dormant—in winter and early spring as well as in late fall. Be sure that the bushes don't grow too dense. Ideally, a well-pruned bush would have about five one-year-old shoots, four two-year-old canes, and three three-year-old canes.

Insect Pests and Diseases

Currants share the same troubles as gooseberries, and these pests and diseases should be dealt with in the same way (see Index). Compared to many fruits, however, both currants and gooseberries are relatively trouble-free.

Propagating Currants

All types of currants can be propagated in the same ways as gooseberries (see Index). Like gooseberries, they do not come true from seed and are only grown in that manner in order to obtain new varieties. When planting them from seed, use the identical method for gooseberries grown from seed.

Currants are more difficult to layer than gooseberries, although they can be increased by layering if the upright shoots can be bent down and pegged to the ground. The bushes are usually propagated by hardwood cuttings. Early in the spring, before growth starts, take 8- to 12-inch-long cuttings from one-year-old canes on a productive bush (or make the cuttings in late autumn and bury them over the winter in moist sand or sawdust in a cool place). Insert each cutting about three quarters of its length into the ground, leaving only two eyes or buds above the ground. The cuttings can be grown 6 inches apart in a nursery bed and transplanted to their permanent place in the garden after 1 to 2 years, or they can be planted in their permanent spot to begin with. Since 50 per cent of currant cuttings ordinarily survive if made properly at the right time, watered sufficiently, and covered with a glass jar the entire first year, two or three cuttings placed in a permanent site should suffice to obtain one that roots and becomes a currant bush.

HARVESTING

Fortunately, currant bushes are thornless and present none of the problems gooseberries do to the picker. One-year-old currant bushes should bear fruit two years after planting and be at the height of their productive powers by the time they reach five years old. They yield up to 6 quarts a bush and go on bearing for as long as thirty years. Currants for eating out of hand or dessert should be dead ripe and picked just before eating, but fruits that are to be used for jam and jelly making are best picked firm and not fully ripe. Make sure that all berries are dry, and do not pick them one by one. Twist the clusters off the branch first and then strip the berries from the clusters.

Currant Favorites

Currants have a high pectin content and are excellent for making jams, actually supplying all the pectin that is needed to make raspberry-currant jelly. But they are used for other tempting dishes as well, some of the best ones following.

General Robert E. Lee's Currant Wine

"Put three pounds of brown sugar to every squeezed gallon of currants. Add a gallon of water, or two, if juice is scarce. It is better to put it in an old wine-cask and let it stand a year before you draw it off." (Copied from a recipe in Mrs. Lee's handwriting)

Currant Pie

Mix 1½ cups sugar and 2 tablespoons flour and sprinkle over 1 quart of half-ripe currants. Line a pie plate with pastry. Fill with the currants and adjust the edges of the top crust, carefully, as currant pie is very juicy. Bake at 450° 10 minutes; then reduce heat to moderate (350°) and bake 30 minutes more or until crust is done.

Currant-Raspberry Preserves

Three pounds each of raspberries, currants, and sugar. Wash the berries. Boil the currants for ½ hour and strain them through a sieve. Simmer the currant juice and sugar in a saucepan for ¼ hour. Add the raspberries, letting the mixture boil up and remain boiling for ten seconds. Remove mixture from heat and pour into jars.

Currant Shrub

For every pint of currant juice, add 1 pound sugar. Boil the juice and sugar together. Stir constantly while it cools, and when cold, bottle it. A refreshing warm-weather drink when mixed with a glass of ice water.

Currant Catsup

Boil together for 1½ hours and then bottle all the following ingredients: 5 pounds currants, 3 pounds sugar, ½ pint vinegar, 1 teaspoon cloves, 1 teaspoon cinnamon, 1 teaspoon salt, 1 teaspoon allspice, ½ teaspoon black pepper, dash red pepper.

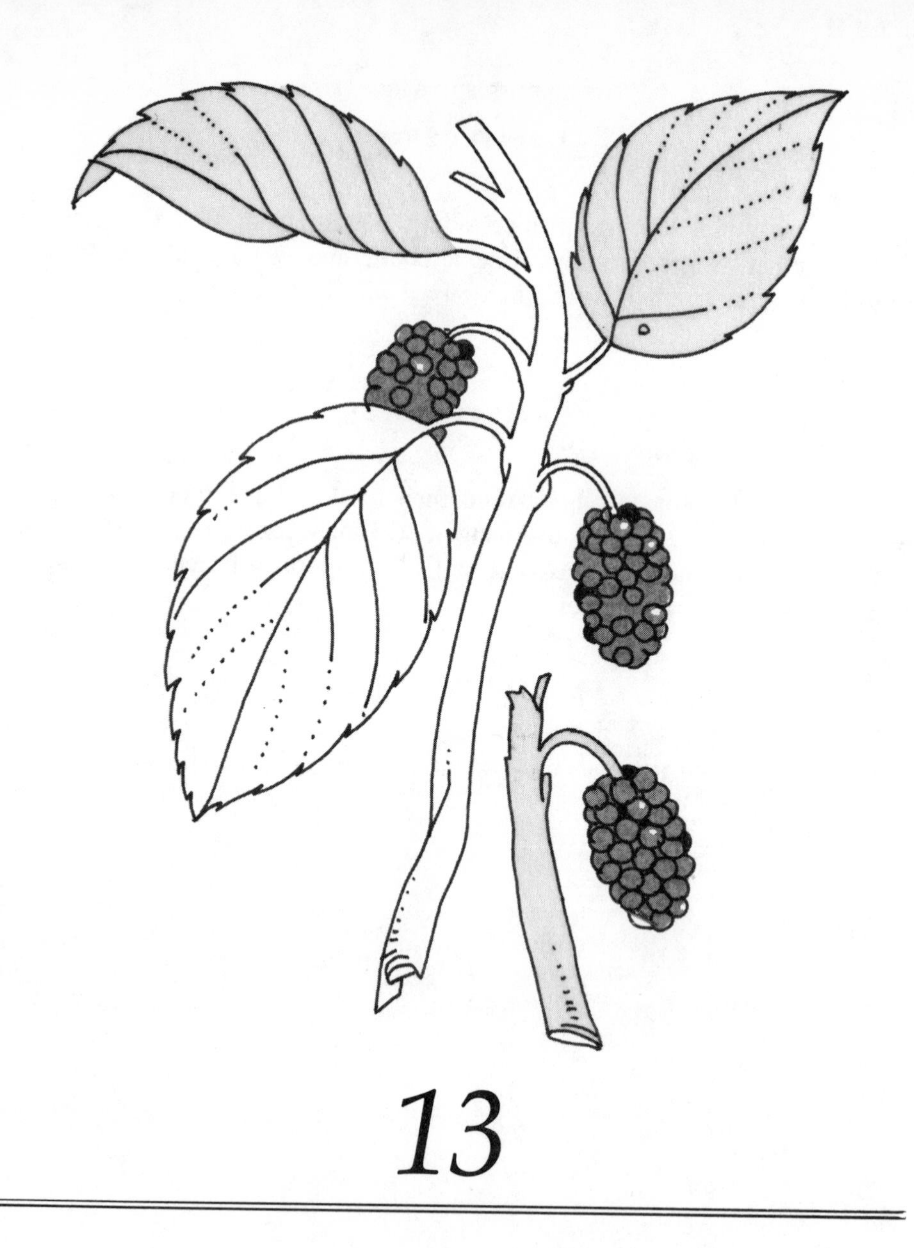

13

Gathering Around the Mulberry Bush

It's hard to understand why a book, or even an essay, hasn't been written about the mulberry, which is a sentimental fruit, the fruit of youth or lost youth to so many, the first food ever gathered with love, and should be mentioned here if only for that reason—in memory of every child who climbed and picked this tenacious tree everywhere from country meadows to littered lots in the most polluted cities.

Who doesn't remember a gnarled, crooked mulberry tree of days past, a child with purple-stained fingers, lips, and teeth cramming sweet fruits into his mouth as fast as he can pick a fistful? And the stained white shirt afterward (no other color shirt was appropriate for mulberry picking)? What other tree belongs to everybody? Unwanted, unheralded, uncultivated in America, the tree of birds and chickens and hogs and boys and girls, whose leaves won't come out until spring is a certainty and die with the slightest breath of frost, the tree of youth, which like the idea of youth seems to hang on with a tenacity cultivated trees have had bred out of them, an escape springing up from seed dropped by birds in the most unlikely places and defeating every attempt by "experts" to eradicate it.

No one would suspect that these orphan survivors in the wild, these "dirty trees," as the experts call them—they stain clothing and discolor the concrete with which we are paving the planet—are among the most aristocratic of trees. Through the ages, and even today, the mulberry has been valued for its leaves (used in silk culture), for its inner bark (used to make vellum paper), and for the wild juiciness of its fruit (far from always being "flat and insipid," as the experts say). It also sees service as a windbreak, as a shade tree, as an ornamental (especially in varieties like the weeping mulberry), and the related "paper mulberry" (*Broussonetia papyrifera*) is cultivated exclusively for paper in Asia. The mulberry is even valuable as a decoy in the war fruit growers have been waging for centuries against birds.

Though it isn't grown commercially in America anymore, no tree has a longer history of cultivation than the mulberry. In his *L'origine des plantes cultivés,* a work painstakingly compiled from ancient writings and archaeological evidences, Alphonse de Candolle wrote that mulberries were one of the 27 crop plants cultivated in the Old World (Europe, Africa, and Asia) more than four thousand years ago—strawberries and raspberries having been raised only half as long by peoplekind.

The mulberry tree has been valued so many years that it is part of ancient fable. An Asiatic legend recounted in Ovid's *Metamorphoses* tells how the berries became red. Pyramus, a Babylonian youth, loved Thisbe, the girl next door, and when their parents forbade them to marry, they exchanged their vows through an opening in the wall that separated their two houses. Thisbe agreed to meet her lover at the foot of a white mulberry tree near the tomb of Ninus outside the city walls, but on reaching their trysting place she was frightened by a lion, dropping her veil when she fled deep into a cave. The lion, its mouth red from another kill, ripped up the veil, covering it with blood, and when Pyramus arrived and found the bloody veil, he thought that Thisbe had been killed and devoured by the beast. Throwing himself on his sword, he committed suicide just as Thisbe emerged from the cave. Distraught at the sight of her dying lover, Thisbe, too, fell upon his sword and committed suicide, the blood of young love mingling and flowing to the roots of the white mulberry, which thereafter bore only red fruit.

Legend also has it that the mulberry takes its botanical name, Morus, from the Greek *moros,* meaning "a fool." This has no connection with the Pyramus myth, according to the *Hortus Anglicus,* but is related to the fact that it "can't be fooled," that the tree "is reputed the wisest of all flowers as it never buds till the cold weather is past and gone." As for the word "mulberry" itself, which should properly be "morberry," it more prosaically derives from the Latin *morus,* which became *mure* in French. The English called the berry the *mureberry* at first, but this was difficult to pronounce (too many "r's") and was eventually corrupted to *mulberry* in everyday speech.

The mulberry undoubtedly owes its high pedigree to the fact that its leaves are the favorite food of silkworms, which produced silk worth more than its weight in gold in Roman times. The Chinese invented sericulture, probably in prehistory, feeding silkworms the tenderest and freshest leaves from the white mulberry (*Morus alba*), something like one ton of this nourishment needed before the voracious worms can produce just twelve pounds of silk. In A.D. 550 silkworm eggs were finally smuggled into Europe by two Persian monks, who concealed them in a hollow cane, and silk manufacture began there in earnest. At first Europeans grew the Persian black mulberry (*Morus nigra*) to obtain food for the silkworms and these trees were planted throughout Europe.

Even the English tried growing mulberry trees in attempts to create a silk industry. They are recorded as having been planted as early as 1538 at

Syon House in England, but a little later James I encouraged the cultivation of mulberry trees and went so far as to have over 100,000 trees planted in the eastern and midland countries, some of these gnarled and crooked trees, almost 500 years old, still alive today. James's effort failed, as would others in the future, and though "mulberry gardens were common in the neighborhood of London . . . either from the climate, or the prejudices of the people, the growth of silk never prospered."

James I's efforts to introduce silk culture to America met with a similar fate. The first shipment of mulberry trees here in 1609 was destroyed in a shipwreck. About ten years later a shipment of trees did get through and the cultivation of silk began in Virginia, although over the years it was to prove as disastrous as the attempt Cortez made to establish a Mexican silk industry in 1522. American colonists were encouraged, however; they were even treated to rhapsodical rhymes like the following: *Where Wormes and Food doe naturally abound/A gallant Silken trade must there be found./Virginia excells the World in both—/Envie nor malice can gain say this troth!*

Later, Benjamin Franklin tried to establish a silk industry in Philadelphia, and many states, including Connecticut and California, have since offered bounties for the growing of mulberry trees. Silk growing has never prospered in America, mainly because of the problem of competition with cheap oriental labor, but there have been schemes aplenty. One of the most memorable was the "Mulberry Mania" of the 1830s. Many will recall the earlier European "Tulipomania," when speculators sold tulip bulbs for as much as $5,000 apiece, but few recall our native mulberry mania. This was a craze for planting the Philippine white mulberry variety *Morus multicaulis* ("many stemmed") in expectation of making great profits in the silk industry. The leaves of these trees were said to be superior to all others for silkworm feeding and millions of them were planted in the "multicaulis fever" that ensued.

The "fever" began in Connecticut, where the seven Cheney brothers founded America's first silk mill at South Manchester in 1838, after having experimented with silk culture for five years. One year, from 300 mulberry trees laid horizontally in the ground, there sprang 3,700 shoots, enough to feed 6,000 silkworms. This meant bushels of cocoons and yards of much-wanted silk. Many farmers in Connecticut, New Jersey, Pennsylvania, and Ohio followed the Cheneys' example and turned their fields into nurseries. Across America books and articles were published about raising mulberry trees. Silk societies were formed, bounties offered. Prices spiraled crazily. In 1838, 2½-foot cuttings skyrocketed from $25 to $500 per hundred. In Pennsylvania alone as much as $300,000 changed hands for mulberry trees in a week and trees were frequently resold by speculators at great profits within a few days. But by 1840, mulberry trees glutted the market at 5¢ each and after the "mulberry blight" struck in 1844, whole groves of trees dying, the "Mulberry Mania" finally came to an end. When speculation collapsed and the costly so-called "golden-rooted trees" were uprooted from plantations in 1839, disgruntled investors coined a new word that served as slang of the time: "multicaulised," meaning "run out, good for nothing, disliked." The only mulberry tree speculation that rivaled the "Mulberry Mania" would come almost a century later when a promoter conned investors into buying mulberry trees by promising that he could guarantee "a precolored silk" by providing "rainbow hued silkworms" that had actually been injected with dyestuffs.

Silk culture hasn't been the only reason mulberry trees have been grown throughout the world. As noted, the inner bark of the paper mulberry (*Broussonetia papyrifera*), an ornamental member of the mulberry family, with its curiously gnarled and twisted trunks in older trees, has long been used by the Japanese for making paper (from which paper umbrellas are still fashioned), and until relatively recent times this species provided tapa cloth, the chief clothing of the Polynesians. In 1751 Thomas Jefferson experimented with the paper mulberry in America. Today the paper lanterns of the renowned Japanese sculptor Noguchi, which are designed not only to contain but *become* light, are made from a mulberry-bark paper.

Mulberry fruit has been a favorite wherever any species of the tree has been grown. The Chinese considered the mulberry a delicious, useful fruit, employing it in many potions and sauces as an aphrodisiac for "the improvement of virility"—including a dish of shark stuffed with mulberries. In England the Saxons filled gold and silver goblets with the richest "morat," a drink made of honey flavored with the juice of mulberries. *Turner's Herbal* (1562) advises that "the juice of the rype mulberries is a good mouth medicine" and a later manual claims that "mulberries are grateful, cooling and astringent."

Far from being an unknown fruit, as it is today, the mulberry was much celebrated in literature. Considered a harbinger of spring—"Whensoever you see the Mulberie begin to spring, you may be sure winter is at an ende" —the tree became part of ancient song and story. In the *Seven Champions,* Eglantine, daughter of the King of Thessaly, was transformed into a mulberry tree. Shakespeare mentions mulberries in both *Venus and Adonis* and *A Midsummer Night's Dream,* the last play making the Pyramus and Thisbe myth the subject of a "tedious brief scene" of "very tragical mirth" presented by Nick Bottom and his rustic players, which is a travesty of the beautiful legend. On the other hand, no one knows how old is the perennial English-American game where children hold hands and sing: "Here we go round the mulberry bush,/The mulberry bush, the mulberry bush/Here we go round the mulberry bush/On a cold and frosty morning."

It's even said that Ludovic Sforza, patron of Leonardo da Vinci and one of the most powerful and unscrupulous princes of the Renaissance, called himself *il Moro* after the mulberry, because he prided himself on his prudence, which he felt equalled the mulberry tree's caution in leafing out each spring. Sforza could just as well have been called *"il Moro"* because he was as swarthy as a Moor, but in any case, the mulberry was well-known enough without him. There was a place called Mulberry Gardens in London from early times and America had Mulberry Streets and Mulberry Corners as far back as colonial days. Here, in addition to the white and black mulberries imported from Europe, we have our native red mulberry, *Morus rubra,* which has not only been eaten out of hand and used in recipes for syrups, wines, jams, tarts, and other desserts, but has been considered an important food for poultry and pigs in the South. So ubiquitous is the red mulberry in America that it was often used as a synonym for "raspberries" in colonial times. Longfellow wrote of mulberry trees in one of his poems, and well into the twentieth century ripe mulberries ready to be picked were a common sight just a few yards from main streets in cities. Many writers after Longfellow praised mulberries. D. H. Lawrence hailed mulberries; in fact, he liked to take off his clothes and climb into mulberry trees to meditate. Probably the best proof of the berry's popularity is the fact that it was chosen as the code name for the engineering feat of installing prefabricated harbors off the coast of Normandy prior to the landing there by Allied Forces in World War II, "Operation Mulberry" making the supply of Allied Forces far more efficient, and hastening the end of World War II.

Mulberries are not grown commercially anywhere in America today—not even for silk culture—and anyone wanting to taste the delightful berries with the long pedigree must either gather or grow them, for the fruit is far too perishable to be carried even by posh fruit stores. Red, black, and white kinds grow wild in lots and woods everywhere, including some that were among the earliest trees planted in America, and a number of good varieties are still available from nurseries.

Gathering Mulberries

In case you'd like to gather berries, seedlings, or cuttings from the wild, here is a list identifying a number of mulberry species. All have alternate sharply toothed, heart-shaped leaves and small, greenish flowers in stalked hanging catkins, the male and female separate. The fruits, resembling blackberries or boysenberries, are technically "aggregate fruits," consisting of a dry fruit (an "achene") covered with the fleshy sepals from several flowers:

Red or American Mulberry (*Morus rubra*). The only native North American mulberry grows to a height of about 70 feet and is spreading in habit. Most common of all mulberries growing wild here, it has a scaly brown bark and ovalish leaves, hairy on the underside, that are 4 to 6 inches long and sharply toothed. Its male flowers are in slender catkinlike spikes about 2 inches long and the female catkins are about half this size. The red mulberry's pleasant, piquant fruit, from 1 to 1½ inches long, is cylindrical and light red but turns to a purplish red when fully ripe.

WHITE MULBERRY (*Morus alba*). The mulberry of the silk growers, native to China, is low-branched and grows from 30 to 50 feet tall, its leaves 3 to 5 inches long and its female flowers in catkins from ⅓ inch to 1 inch long, the male spikes slightly longer. White mulberry fruit can range from ½ to 1¾ inches long, and is a white or pinkish violet in color. The fruit is sweet and insipid in taste, not really good eating, but the variety *Morus alba tatarica* (the Russian mulberry), a smaller tree that is often planted as a windbreak, has good, red-colored fruit. A form of the Russian mulberry frequently seen in America is the weeping mulberry developed toward the end of the last century in Carthage, Illinois, and long one of the most popular ornamentals for the front lawn in the country. *Morus multicaulis,* the Philippine tree which inspired our American "Mulberry Mania," is another variety of the white mulberry, a much smaller, shrublike tree with larger leaves that isn't much valued for its fruit.

KOREAN MULBERRY (*Morus acidosa*). This bushy, shrublike species grows only ten feet or so tall and has juicy, sweet, dark red fruit. Its male catkins are 1¼ inch long while the female catkins measure only about ¾ of an inch. Not many of these are found growing wild, but it is an excellent tree to plant, especially where space is a problem, if you can find a nursery that offers it.

MONGOLIAN MULBERRY (*Morus mongolica*). Similar to the Korean mulberry above, but not as valuable, it grows up to 25 feet tall and its fruit is pale red, sweet, not as juicy as its Korean relative.

BLACK MULBERRY (*Morus nigra*). Most of the large-fruited varieties developed by nurserymen derive from this Persian species, which is widely planted in Europe and America. The black mulberry grows 20 to 30 feet tall, lives hundreds of years in many instances, and in its old age is a wide-spreading, large-headed tree with beautiful gnarled and crooked branches. The male flowers are in slender catkins ½ to 1 inch long, while the female catkins are thicker and about half as long. The fruit, which is much larger in nursery varieties, grows up to 1 inch long. When ripe it is purplish black with a slightly acidic flavor, but when allowed to ripen on the tree a few weeks longer turns almost fully black and is sweet and delicious eaten out of hand. It is said that a variety of black mulberry grown as a fruit tree in Iraq and Turkestan produces very large seedless fruits, but to my knowledge no western nursery offers this type.

THE BEST MULBERRY VARIETIES TO GROW

The mulberry genus, *Morus,* belongs to the same family as the fig, breadfruit, and rubber trees, not to mention *Cannabis sativa,* the plant from which marijuana is obtained, and the genus *Humulus,* from which the hops used for making beer are obtained. There are about twelve species of the *Morus* genus, but only five or so are planted for their fruit, many varieties offered by nurseries among these. Some varieties are everbearing—stretching their fruiting over the entire summer—while others ripen early and still others relatively late in summer. Nurserymen can recommend varieties suited to your area and ripening times, so that you can pick mulberries over a four- or five-month period. Some of the better ones are King, Downing, Wellington, Black English, Russian, Thorbum, Troubridge, Shaheni Red, Everbearing, Hicks, New American, Bideneh Seedless White, and Black Persian (especially good for the West Coast and southern regions). All should bear a year or two after planting. Sources include the New York State Fruit Testing Cooperative, whose address is given in Appendix III. The North American Fruit Explorers (1848 Jennings Drive, Madisonville, Kentucky 42431) is not a nursery, but should be able to provide information on where to obtain Persian varieties like King, Shaheni, and Bideneh.

When, Where, And How To Plant

Mulberry trees ordered bareroot from a nursery or propagated from the wild are best planted in early spring when the ground begins to warm up. The trees should be set about 30 feet apart. They prefer a rich, loamy soil that is somewhat moist, but they are not at all finicky and will grow almost anywhere, even in sandy soil. In cold or very wet areas, mulberry fruits tend to drop off the tree before fully ripening, however, and trees grown in cultivated ground bear better than those planted on a lawn. Needless to say, full sun will ensure more and better fruit, yet all varieties will produce berries in fairly deep shade as well. In other words, you can just plant a mulberry tree and forget about it—it will usually make it on its own—but for the best crops of berries, give the tree a little care. If the soil is poor where you plant a mulberry, for example, dig a deep hole and fill it with compost before transplanting. If the ground is dry, water the tree through the summer until its roots get established. And if the tree is planted in a cold climate, try to place it against a wall in the warmest part of the garden.

Space-saving Mulberries for the Garden, Patio, and Greenhouse

The ten-foot-high Korean mulberry species and other bushlike types are sometimes planted to save space if the garden isn't big enough to accommodate a standard tree. Or almost any variety of mulberry can be trained into space-saving forms. One method is simply to keep a standard tree small by constant pruning, restricting it to about a three-foot stem and a small head. A handsome *wall tree* is another alternative—just train the mulberry in the shape of a fan against a south wall, keeping the branches a foot apart and cutting back all side shoots to six leaves in the summer to form fruiting spurs. Feed wall trees with a weak solution of liquid manure when they begin to fruit heavily.

The ultimate in space-saving mulberries is probably the *pot tree,* which is grown in a twelve-inch pot on the patio or in the greenhouse. It can be simply a bush form or might be trained in a unique pyramid shape. Pot trees should be cared for like any standard tree, but they do need to be repotted every year in early spring before new growth begins. Keep pot trees from getting too tall by nipping their leaders when necessary and cut back the laterals to six leaves in early summer to encourage the production of fruit-bearing spurs. They, too, should be fertilized with a weak "manure-tea" (manure and water) solution when fruiting heavily.

Assorted Mulberry Ploys

Birds are so partial to mulberries that many home gardeners and commercial fruit growers plant a few mulberry trees to act as decoys and distract our feathered friends from blueberries and strawberries.

Farmers often plant mulberries to satisfy their chickens, which eat their fill of mulberries and don't bother other berries as they patrol the garden for insects.

In the past, farmers who raised pigs planted a mulberry area with oaks, or planted mulberries between standing oaks, to provide their porkers with acorns and mulberries, two of their favorite foods.

Care Of Mulberries

Watering, Fertilizing, and Pruning

The mulberry is a hardy tree that thrives throughout the United States, growing wild in the southern Appalachian forest region, and a number of varieties can survive temperatures as low as $-20°$ F. when established. Aside from watering an established tree during dry spells and turning over the soil in spring and autumn where it is not grown on a lawn, it needs little care. Fertilization is not necessary on standard trees, although a thick mulch of compost or leaves around a tree wouldn't hurt it and would keep the soil moist as well. As for pruning, standard trees need only be pruned of dead branches that are causing overcrowding. These should be removed or thinned out in the winter when the tree isn't growing and cut flush to the trunk, the wounds covered with tree-wound paint. The mulberry is a weak-wooded tree and loses many branches to the wind or from kids climbing in it, but its fast-growing habit compensates for this.

Cross-pollination

Mulberry trees purchased from a nursery do not need cross-pollination; that is, there will be male and female flowers on the same tree. But some trees, especially those gathered in the wild, will be dioecious, which means that only either male or female parts of the flower are on a single tree. Nothing can be done about this if the tree is a male except to plant a female tree about thirty feet from it. If the tree has just female parts, however, there are several solutions. There will likely be a male tree in the locality that will fertilize it, or, in the unlikely event that there isn't one, a male tree can be planted. Or else simply cut sprays of male bloom from a tree a great distance away, put them in bottles filled with water tied to the female tree and let the insects do the rest.

Controlling Pests and Diseases

Though it is remarkably free of diseases and pests, the mulberry does have a few enemies, most of which can be easily controlled:

SPRUCE BUGWORM. These worms, which infect evergreens as well, are best controlled by handpicking the cocoons containing their eggs and burning them. A commercial preparation called Trick-o is also effective, and commercial chemical sprays are available, too.

RED SPIDER MITES. Chemical sprays aren't needed here. There are several safe, organic sprays like rotenone that will control red spider mites, which sometimes damage mulberry leaves. Forcibly spraying the leaves with water, taking care to wash the undersides, will also help some. So will a 3 per cent oil spray or onion juice spray, and the dust pyrethrum.

FALL WEBWORMS. Moth larvae that live in colonies, each under a web, and feed on foliage. Prune any branch tips containing them to control the worms.

CANKER DISEASE. Entire branches can be lost to this disease, which is caused by a fungus. The only control is to remove all infected branches in the fall or winter, cutting them a foot from any diseased portion and burning them.

MILDEW. In southern states mildew occasionally mars the foliage of mulberries, but it is not a serious disease and seldom warrants spraying with a commercial chemical preparation.

POPCORN DISEASE. A disease that is an oddity more than a menace, popcorn disease swells the size of some of the mulberry fruits, doing very little damage.

PROPAGATION

Mulberries are extremely easy to grow and you can save money propagating them rather than buying trees from a nursery. Either select a tree with tasty fruit growing in the wild or impose upon a neighbor who has a good named variety and try one of the following methods.

SEED PROPAGATION. Wash seed from ripe berries in late summer and store it, thoroughly dried, in a cool place. Plant the seed in a coldframe in March, or out in the open ground where you want the tree to grow in late April or early May, whenever spring is there to stay in your area. Mulberry trees rarely fail to grow from seed, but there are two decided disadvantages that make the practice inadvisable: mulberries do not come true from seed and they take many years to bear fruit when grown in this way.

If you see a mulberry tree coming up from seed dropped by birds in an out-of-the-way place, however, you might give it a chance and let it grow. Mulberries will sprout up almost anywhere—in the midst of privet bushes, against the foundation of houses, out of cracks in concrete sidewalks, and even out of the roots of other trees. I have a favorite reaching for the sun on a forty-five-degree angle out of the roots of a pignut hickory. It must be about seven years old, though I first noticed it a few years ago, and it is just beginning to bear small but tasty berries.

SMALL CUTTINGS. Well over 50 per cent of small cuttings taken from mulberry trees will root and produce trees identical in fruit and other characteristics. It is best to make cuttings in early autumn from the shoots of the current year's growth, though the cuttings can be taken in early spring as well. Mulberry cuttings should be about 1 inch long and have a heel of 2-year-old wood about 3 inches long attached. Plant them in a shady spot with just two bud eyes showing above the ground. In a year roots will have formed and the cuttings can be transplanted to their permanent location.

LARGE CUTTINGS. This is the quickest way to propagate a mulberry tree, though it is not as reliable as using small cuttings. However, even large mulberry branches will root if inserted deeply into the ground and protected during the winter. In autumn cut off a branch about 5 feet long, removing all subsidiary branches, and insert it about 18 inches into the ground where you want the tree to grow. Tie the branch to two stakes to keep it steady and upright and cover it with a tarpaulin or other protection for the winter. By the next autumn you should have a fruit-bearing mulberry tree or bush.

Make four or five large cuttings this way to increase your chances of one rooting.

GROUND LAYERING. Another very easy method. In late summer, early autumn, or even spring, a strong young mulberry branch still growing on a mother tree is cut shallowly on its underside and bent down to the ground, where it is buried in a slit trench about 6 inches deep. The branch can be held from springing back out of place by anchoring it with a forked twig or a piece of sturdy wire. Make sure that the branch is bent down as flat as possible and that about 3 inches of the tip protrudes above the surface of the ground. Then close the slit in the ground tightly over the bent branch with a stamp of the foot. The injured branch will root in a year or less and then can be cut from the mother plant and transplanted to its permanent location.

AIR LAYERING. Here a young branch still on the parent plant is injured (slit or debarked) 10 to 12 inches from its tip. Make the wound about ½ inch wide. Moist (not wet) sphagnum moss is then wrapped tightly around the wound, which can be sprinkled with rooting hormone, and a covering of polyethylene plastic is wrapped tightly around the moss, this taped or tied closed at both ends. Polyethylene plastic permits air to enter but confines humidity. It is very important that the tie on the plastic be tight, however, as any rain that enters will wet the moss too much and prevent rooting. Air layers made in late spring will root by late fall and can then be severed from the mother plant and transplanted whenever you want them, or left on the plant until spring.

GRAFTING. In addition to propagating mulberry trees you can plant seed as previously noted to obtain sound native stock and graft cuttings from higher-yielding or larger-berried trees onto the stock. Usually this is a job for an experienced nurseryman, but if you want to try, graft or shield-bud the stocks by the general methods outlined in any gardening encyclopedia.

MULBERRY HARVESTING

If you want to eat mulberries, you'll have to stay ahead of the birds. Either pick the berries every day to forestall them, net the trees, or try hanging up aluminum reflectors and noise-making devices to scare them off (see Appendix I for more possibilities). Netting mulberries is probably the best solution, but this is only practical on smaller trees. Larger trees produce so many berries over so long a period of time, though, that daily picking of the ripe fruit assures plenty for both you and the birds. Bush trees and dwarfs can, of course, be picked by hand, while an easy way to pick large trees without using a ladder is to spread a sheet under the branches and then give the tree a few gentle shakes. This last method, used for centuries by Italian mulberry growers, shakes ripe berries to the ground while unripe ones remain clinging to the branches.

Using Mulberries

Mulberries do not keep very well and must be used within a few days, the sooner the better. The berries have a sugar content of about 9 per cent. Besides eating them out of hand, or serving them fresh as a dessert for supper with a little citrus juice sprinkled over them, they can be preserved and used in numerous dishes. Dried mulberries are used like raisins in cookies and muffins, while the fresh berries are part of many recipes for tarts, pies, jams, jellies, wine, and other beverages. Here are just a few possibilities.

Mulberry Jam

Wash the berries carefully, drain, and remove the caps and stems. To each pound of prepared fruit allow an equal weight sugar. Crush the berries and bring slowly to a boil, stirring constantly. Add the sugar and boil until the fruit mixture has thickened to jellylike consistency. Stir throughout the cooking. Pour into hot sterilized jars and seal. If the seeds in mulberries are objectionable, boil the fruit for a few minutes, then put through a fine sieve to remove the seeds before weighing the fruit and adding the sugar. This is a real delicacy, mulberry jam from the South of France selling for $7.50 a one-pound jar when available.

Mulberry Muffins

¼ cup shortening
⅓ cup sugar
2 beaten eggs
2 cups flour
5 teaspoons baking powder
1 teaspoon salt
⅔ cup milk
½ cup mulberries

Preheat oven to 400°. Cream shortening and sugar together. Add eggs and mix well. Sift 1½ cups flour, baking powder, and salt together. Add this mixture to egg-and-sugar mixture alternately with milk. Sprinkle mulberries with remaining flour and stir in lightly. Bake in greased muffin pans for 25 to 30 minutes.

Mulberry Mousse

3 cups sugar
1½ quarts water
7½ cups mulberries, crushed
6 tablespoons lemon juice
3 cups heavy cream

Boil sugar with 1½ quarts water 8 minutes. Force mulberry pulp and juice through a sieve and add to syrup. Add lemon juice. Blend thoroughly. Place in a freezing tray. Stir 3 times at 30-minute intervals. Remove and fold in cream, beating 2 or 3 minutes if crystals are large. Return to freezing tray 2 to 3 hours longer.

Rummy Mulberries

Dissolve sugar in potent rum, ½ cup sugar for each cup of rum used. Wash and drain mulberries and place in jars, covering with the rum-sugar solution. Tightly seal and store in a dark closet for three months before using. Especially good served over ice cream.

14

Unusual Berries:
From Candle and Soap Makers to Bathtub Gin and Afterdinner Mints

Berries you can work magic with . . . berries to stun fish with . . . berries that are supremely edible . . . others that can be used to make candy, raisins, soft drinks, lemonades, coffee, health teas, laxatives, dyes, heart medicines, or even soap, candles, cosmetics, and bathtub gin . . . All of the following are uncommon berries that can be grown in the home garden. Although several require special conditions, most gardeners will be able to find the right place for at least a few of these little-known fruits. Some sources are indicated, others may be located in references like the *Plant Buyer's Guide* published by the Massachusetts Horticultural Society, and still others of these plants will only be found in the wild.

APPLEBERRY (*Billardiera longiflora*). Not much grown in America, except as an ornamental in the South, this Australian berry has for nearly two centuries been commonly raised in England, where it has proved hardy outside in the milder parts of the British Isles and is grown in the greenhouse in cooler climates. Named botanically in honor of Jacques Julien Labillardière, a celebrated French botanist and traveler who collected it, the appleberry is an unusual, tasty, blue berry that is only about an inch long but somewhat resembles an apple in shape. An evergreen climber that will twine up to 5 feet on a south wall or other support, it has long, greenish-yellow, bell-like flowers that change to purple. The variety *B. longiflors fructo alba* has white berries, and another species *B. scandens* climbs higher and bears fruit later. The appleberry prefers an acid soil with shade at its feet and should be planted in early spring in a compost of loam, leaf mold, and acid peat in equal proportions, with good drainage. Pruning consists only of removing dead wood in the early spring, and fertilizing is not necessary. The plant roots readily from half-ripe wood cuttings taken in late July and inserted in sandy soil. It can also be grown from seed under the same conditions. Appleberries bear profusely and the fruit makes a delicious dessert eaten raw out of hand.

BARBERRY (*Berberis vulgaris*). Few gardeners would believe that the common spiny barberry, an ornamental bush often planted for hedges in place of privet and valued for its flowers, its gorgeous fall foliage, and its striking red berries that often hang on the bush all winter long, provides fruit that can be eaten in a variety of ways. The English, in fact, once cultivated the handsome bush for its berries, and the Arabs before them grew barberries for their sherbets. The barberry, in fact, may take its name from the Arabic name for the fruit, *berberys,* a shell, possibly in reference to its leaves being hollow like shells, but in any event it is associated with the Berbers, who cultivated it on Africa's Barbary Coast. Some fifty species are widely grown in America, most frequently *Berberis vulgaris,* the common or European barberry. The deciduous European barberry would be even more commonly grown here if it weren't the host for a serious wheat rust that excludes it from wheat-growing regions. Evergreen species of the genus, however, are not carriers of this fungus.

The European barberry long ago became naturalized in America. Thoreau in his letters mentions going off "a barberrying" and the acid berries were gathered by the earliest settlers for use in preserves, pies, and sauces. I found it interesting to learn that this berry, which many avoid today as if it were poisonous, can even be eaten green, pickled in vinegar, and used like capers as accompaniment to cold meats. Mrs. Glasse in her famous cookbook mentions "a garnish of barberries and lemon" and apparently the green leaves of the thorny bush were used to make a "sauce to eate with meates" in the sixteenth century. Barberry roots and bark yield a bright yellow dye and a medicine that colonists valued. The berries, which contain a great deal of citric and malic acid when ripe, have also been used as a lemon-juice substitute in cold drinks, to make wine, for flavoring punches, and, in India, as a dried dessert raisin.

All barberries are simple to grow, requiring no fertilizing and little or no pruning except for removing weak old branches occasionally. They can be planted in sun or shade, do well in ordinary garden soil that isn't waterlogged, and can easily be propagated from seed sown fresh from the berry in autumn, which will germinate in the open by the following spring. All of the following species are suitable for ornamental hedges and as attractive specimen plants as well as for their fruits.

European Barberry (*Berberis vulgaris*). See above. The most fruitful of barberries, but not evergreen, not the most attractive, and a carrier of rust disease. There are red, yellow, violet, black, and white fruited forms. Hardy from zone 2 southward, grows 5 to 10 feet tall or more.

Seedless European Barberry (*Berberis vulgaris* var. *asperna*). A scarce variety of the above that has no seeds and was used in a famous Rouen conserve.

Magellan Barberry (*Berberis buxifolia*). One of the best of the evergreens, hardy from zone 4 southward and grows up to 8 feet high. One form of this species has thorns longer than its leaves and its purplish-black berries are used for a popular preserve in its native Chile.

JAPANESE BARBERRY (*Berberis thunbergi*). More cultivated than almost any other shrub in America, this barberry grows 4 to 6 feet high, has brilliant scarlet autumn foliage and light-red, winter-persisting berries that are used for jelly. Hardy from zone 3 or 4 southward.

RAISIN BARBERRY (*Berberis asiatica* and *Berberis aristata*). Both of these shrubs have yellow flowers and purplish berries, but *aristata* flowers more profusely and grows only to about 6 feet tall, while *asiatica* can grow about 8 feet high. Both are half-hardy plants that can be grown outdoors in the South. Their rather large berries are dried and used to make dessert raisins in their native India. Unlike most barberries, they thrive in a compost of loam, peat, and a little sand.

DARWIN BARBERRY (*Berberis Darwinii*). A very handsome evergreen shrub native to Chile that is hardy from zone 6 southward. Grows up to 8 feet tall and yields edible, dark-purple berries.

MEXICAN BARBERRY (*Berberis haematocarpa*). Large, bright-red berries on a plant that grows to about 8 feet tall and is native to New Mexico.

HIMALAYAN BARBERRY (*Berberis angulosa*). Perhaps the largest berries (purplish black) of all on a bush native to the Himalayas and growing up to 8 feet tall.

Other species well worth growing as ornamentals and for their berries are the *wintergreen barberry* (*Berberis julianae*), the hardiest of the evergreens, which can be grown outdoors from zone 4 southward; and *Berberis Wilsonae,* with its coral-red or salmon-colored berries. There are over fifty commonly grown varieties to choose from and improve upon by natural selection.

In case you want to make use of those barberry bushes outside, here is an unusual (untested) recipe for "Candied Barberries" that I came across in an old cookbook:

"Take some large barberries very ripe and of a fine red colour. Leave them in clusters. For 2 pounds of berries cook 2½ pounds of sugar to 'the large feather' [232° F.]. Put in the barberries and boil very hastily to produce 10 to 12 bubbles. Take off the stove. When the fruit is beginning to cool, put it in a hot cupboard leaving it to drain on a cloth until next day. Put it on sheets of paper to drain further. Dust the clusters of berries with fine sugar rubbed through a drum sieve [a very fine sieve]. Put them to dry in a hot cupboard."

BAYBERRY (*Myrica pennsylvanica,* often called *Myrica carolinensis*). You can make your own candles from these berries. In fact, in colonial times, September 15 was known as Bayberry Day, a time "when old and young sallied forth with pail and basket, each eager to secure his share in the gift of nature." For although the bayberry isn't good to eat, beyond the use of its root bark as a medicine, the candleberry, as it was also called, was considered to be "light on a bush," its small, highly aromatic, gray berries yielding wax for candles. Fragrant bayberry candles were made by boiling the waxy bayberries to a thick green consistency and dipping cords into the mixture, the stemless, wax-coated berries often yielding five pounds of wax to the bushel. Besides being used for Yule candles, the wax was employed in making scented soap in place of animal-fat tallow, and in making sealing wax. The berries, which persist on the bush through the winter, were sold for winter bouquets.

The bayberry, 3 to 8 feet high and deciduous, is hardy from zone 2 southward. An easy shrub to grow, it prefers dry sandy soils, where it grows naturally. The same applies to another species used for making candles, the wax myrtle (*Myrica cerifera*), a tall shrub or small tree that grows up to 35 feet high, has evergreen leaves, and is hardy from zone 4 southward. *Myrica Gale,* or the Sweet Gale, still another candlemaking plant, is also called the bog or moor myrtle and should be grown in acid sites in the bog garden (see Cranberries, in the Index). A deciduous plant, it is hardy from zone 4 northward. *Myrica californica,* the California myrtle, also bears berries that can be used for candles, on an evergreen, 30 to 40-foot, hardy shrub. All of these species are easy to propagate by seeds, cuttings, and divisions.

No molds are needed to make your own bayberry or myrtle candles. Simply boil the berries in water until the mixture gets thick. After straining out the seeds and skins, let the wax cool, reheat it, and keep dipping a piece of wick into the hot wax. Pull the wick out, let the wax dry, and dip it in the hot wax again, repeating the process until the wax on the wick builds up to the size of a candle.

BEARBERRY (*Arctostaphylos uva-ursi*). *Arctostaphylos* is the Greek for "bear grape," applied to this genus because bears eat the berries of some species, especially *Arctostaphylos uva-ursi.* Various American Indian tribes enjoyed this mealy berry, too, and made a beverage from it as well. They also used the "kinnikinick's" leaves like tobacco and as a drug. Bearberries haven't been a very popular fruit over the years, even though a tea made from its leaves was long a folk medicine for urinary troubles, but it can be used like cranberries in many recipes—in fact, another name for it is the *hog cranberry*. Bearberry has been called the "prettiest, sturdiest, most reliable groundcover" for poor soils and will thrive in pure sand. A shade-tolerant evergreen creeper with stems up to 6 feet long, it has handsome, leathery green leaves that turn red in the autumn, white pink-tipped flowers shaped like urns, and scarlet berries ⅓ inch in diameter. The stem roots at the joints, *actually increasing where it is stepped on* because this forces the stems into contact with the soil. The care-free plant shouldn't be fertilized and only needs watering in very dry seasons. Its chief disadvantage is that it is extremely difficult to propagate; in fact, the accepted way is to dig up frozen clumps in midwinter and plant them immediately in a prepared bed of 6 parts sand and 4 parts acid humus. Plants bought from nurseries in the spring should be planted in the same manner. Bigberry Manzanita (*Arctostaphylos glauca*) is one of many tree or shrub forms of the bearberry. This California native grows to about 14 feet and has brownish berries about ¾ inch in diameter. Bearberry plants are available from Alpenglow Gardens, 13328 King George Highway, Surrey, B.C., Canada.

BILBERRY (*Vaccinium myrtillus*). "Pinch the maids as blue as Bill-berry," Shakespeare wrote in *The Merry Wives of Windsor.* Bilberries, or whortleberries, or blaeberries, or whinberries, as they are variously called, are closely related to the blueberry (see Index), being of the same genus. The main difference between the two is that the bilberry is a little plant no more than 18 inches tall and usually produces berries singly and not in clusters as cultivated blueberries do. While the bilberry is native to England, there are similar species found in the eastern and northern United States (*Vaccinium pennsylvanicum,* the lowbush blueberry, for one example). Cultivation is exactly the same as for blueberries. In America, incidentally, "bilberry" was never much used as a name for wild blueberries and "whortleberry" was more often applied to the wild huckleberry. In Britain "bilberry" is used to describe several species of blueberries, including *V. uliginosum,* the bog whortleberry or great bilberry. Britons have long gone "a bilberrying" and enjoyed "whorts" or berries in pies, preserves, or eating fresh with cream, but the little plant is much scarcer today than it was in the past. Hard to domesticate, it will only thrive if you provide it with conditions similar to the acid peat soil of the woodland where it lives, the plant tolerating lime even less than other blueberries. Since the plants are partially self-sterile at least two must be planted to ensure cross-pollination.

BUFFALO BERRY (*Shepherdia argenta*). Buffalo meat eaten with the tart scarlet berries of *Shepherdia argenta* was a favorite dish of American Indians, which explains how the berries got the name. Later, settlers learned to make a delicious jelly with the oval berries that was customarily served with a haunch of venison, and the wild bush that bore them began to be planted as a windbreak or hedge on the northern plains. Called also the *Nebraska currant,* the *silverleaf, beef suet tree, rabbit berry,* and *wild oleaster,* it was and is highly valued not only for its berries but because the thorny shrub, 10 to 18 feet high, is among the hardiest in cultivation. *Shepherdia* is named for John Shepherd, a curator of the Liverpool Botanic Garden. At home from the central Plains states all the way up to the coldest parts of zone 1, it will stand dry, rocky soil and windswept sites, but does best in moist loam similar to its native habitat along river banks. The bush also makes a handsome ornamental with its scarlet berries and small, silvery leaves. Its main drawback is its sharp thorns, which makes berrypicking a bit difficult. The smaller *thornless Canadian buffalo berry* (*Shepherdia canadensis*) could be used in its place but has inferior berries.

Buffalo berry bushes are easily started from seed, suckers, or cuttings, and care for them is minimal. They need only be pruned to keep them under control and require no fertilization at all. Since the species is dioecious—bears flowers of only one sex—both male and female plants must be planted in the same area or the bushes won't bear. Berries can be orange or yellow as well as scarlet and their taste improves if the clusters are allowed to hang on the bush until touched by the first frost. They are made into jelly and can be dried to be used like currants, their flavor similar to a cross between a wild grape and a red currant. Bushes are available from Gurney's.

CHINESE CHE (*Cudrania tricuspidata*). Called the *wild mulberry* in China, its leaves used to feed silkworms, this small tree bears perhaps the newest of unusual fruits grown in America. The che grows up to 30 feet tall, has a spreading top with many but not all spiny branches, and can withstand temperatures of −20° F. Its 1½-inch fruits are a maroon color with rich red flesh inside and 3 to 6 small seeds per fruit, ripening in October in the East. Ches are said to bear at an early age and mature trees produce hundreds of pounds of unique, tasty fruit in clusters, on its thornless branches. The tree is dioecious, bearing flowers of only one sex, so male and female trees must be planted in the same vicinity to ensure fruiting. Dr. George M. Darrow, former head of the USDA Small Fruits Division, has pioneered in the development of this new fruit at Olallie Farm, Glendale, Maryland 20769. Another source is Blandy Experimental Farm, Bogen, Virginia 22620. With a respected authority like Dr. Darrow behind it, this would seem an excellent unusual fruit to try.

CLOUDBERRY (*Rubus chamaemorus*). Though it has also been called the *ground mulberry,* the cloudberry is a member of the raspberry family (see Index). This European native bears juicy, amber-colored fruits that taste something like apricots. They are delicious eaten fresh or as preserves, but are rarely grown in the home garden because the plant needs lots of space for its long underground rhizomes to grow and requires cool, boggy conditions to do well. The cloudberry is known as *maroshka* in Russia, in Scotland as *avrons,* and in Norway as *molteberry*. It is easily propagated by seed or by planting parts of the underground rhizomes but is best gathered in the wild.

COFFEE BERRY (*Coffea* species). Coffee beans are actually the seeds of the fleshy berries that grow on the coffee tree, each berry containing two seeds or beans. The plant probably takes its name from Kaffa, Ethiopia, where it originated, but today is principally grown in South America, where it was brought by Europeans in the early eighteenth century; plants from France's Louis XV's greenhouses are the ancestors of all South American plants. It can be cultivated outside in the United States only in the southernmost parts of Florida (zone 9 and possibly part of zone 8) and even these temperatures aren't really hot enough for it to bear well, nor is the altitude high enough. Nevertheless, it is grown as a striking ornamental in Florida and as a greenhouse curiosity in all parts of the country. Two species are usually grown: *Coffea arabica,* common or Arabian coffee, and *Coffea liberica,* Liberian coffee. These are very similar evergreen shrubs about 10 to 15 feet high with many stems, waxlike leaves, and fragrant large white flowers, the first species bearing clusters of cherrylike red fruit and the second black fruit. The showy shrubs yield when only 3 years old, bear for up to 50 years, and can stand severe pruning to keep them manageable in a greenhouse, making them a worthwhile plant to grow under glass. They should be planted in a soil of one part peat, one part sand, and two parts loam, kept moist, and raised in a warm (70° to 75° F.) house. The easiest way to prune them is by the single-stem method, removing all upright branches and allowing the central stems to grow only 4 feet high, which encourages lateral branching. Propagation is usually by cuttings taken from upright branches, or by seed sown in sandy soil during early spring with the greenhouse temperature at 85° F. If you feel this is too much trouble to grow your own coffee beans, try the Kentucky Coffee Tree (see Index).

CORNELIAN CHERRY (*Cornus mas*). The berries of several dogwood species are edible, but those of the Cornelian cherry have the longest history as a food. Called *cornet plums* in England this tree's scarlet berries were once used to make preserves, tarts, drinks, and were even packed in brine to be used like olives. Additionally, the Cornelian cherry has exceedingly hard wood that gives more heat than most firewoods and is said to have been used to build the Trojan horse of antiquity. Its bark yields the red dye used for the traditional Turkish fez and the berries of a dwarf form of it (*Cornus sueica*) were believed by the Scottish Highlanders to create appetite, inspiring them to name the plant *Lus-a-chraois,* Gaelic for "plant of gluttony." Like all species of dogwood, the Cornelian cherry's bark is rich in tannin and has been used medicinally (just as the bark of the flowering dogwood *Cornus florida* was used as a substitute for quinine).

A small tree or shrub native to Eurasia, the Cornelian cherry is a handsome ornamental and stands air pollution better than most plants. Its berries are about ⅝ inch long and acid, but make good preserves. The tree's main fault is its slow-growing habit; it can take 10 to 15 years to produce fruit if grown from seed or cuttings. The Cornelian cherry is hardy from zone 3 southward. Like all dogwoods, it prefers acid soil and a sheltered, moist site. I mention it here mainly because some gardeners may have an old specimen on their grounds without knowing the value of its fruit.

CRACKERBERRY (*Gaylussacia baccata*). No, this doesn't taste like or resemble a cracker. In case you've read about it somewhere, it is simply another name for the common huckleberry (see Index), commonly called *crackerberry, crackers,* and *black snaps,* because its seeds crack or snap so between the teeth.

CROWBERRY (*Empetrum nigrum*). An alpine or arctic plant of the northern hemisphere and the Andes that is suitable for rock gardens or moist, shady spots in the garden, the crowberry or *black crowberry* is an evergreen, mat-forming shrub that grows only a foot or so high. Its black berries, about 1/4 inch in diameter, taste like turpentine to some, but have a nice, slightly acid flavor to others. Once crowberries were used as a medicine to prevent scurvy and were made into a beverage with sour milk, not a particularly appetizing thought to me. There are five species of the genus, none varying greatly except for the color of the berries, which can also be red or purple. All need a strongly acid soil of pH4 to 5 and won't stand prolonged summer heat, for they extend in range to the Arctic Circle. They can be propagated by midsummer cuttings inserted in sandy peat in a coldframe.

FUCHSIA BERRY (various *Fuchsia* species). The color fuchsia, a vivid bluish or purplish red, actually takes its name from the ornamental fuchsia shrubs that honor Leonhard Fuchs, a sixteenth-century physician and botanist who wrote a noted herbal of medicinal and edible plants. The *Fuchsia* genus contains some 100 species, principally of Mexican and South American origin, and can have purple, red, yellow, or white flowers. Not many of these fuchsias are valued for their fruit—especially not the showy types that have been bred for their beautiful flowers over the years—but several species do have interesting, edible berries. One such is the Peruvian shrub *Fuchsia corymbiflora,* a climber that grows up to 6 feet tall, has showy scarlet flowers, and bears purplish berries that resemble figs in taste. In all but the southernmost areas of the United States this species must be grown in the greenhouse, its culture the same as for any fuchsia. Another species, the New Zealand native *Fuchsia excorticata,* can be grown outdoors from zone 4 southward. An unusual, ornamental shrub that produces flowers on its trunk right down to the ground and has a purplish sheen when in bloom because its flower's pollen is bright blue, this species bears subacid, purple-black berries. Its blue pollen was used by Maori girls as facial makeup.

HAWTHORN BERRY (*Crataegus* species). The popular hawthorn or *thornapple,* often grown as a lawn specimen, yields colorful berries that are excellent for making jams and jellies. Hawthorn berries are also recommended as a heart medicine by homeopathic physicians; in fact, prescription tablets made from the berries of *Crataegus oxycantha,* the English hawthorn, are available in the United States and are claimed by some to both treat and prevent heart disease. Many medicines made from the berries can be obtained in Europe. The best hawthorns for jams are two new varieties available from wholesaler Edward H. Scanlon & Associates, 7621 Lewis Road, Olmsted Falls, Ohio 44138 (your local nurseryman can order them for you).

CRATAEGUS PINNATIFIDA MAJOR. A variety of the *Chinese big leaf hawthorn* that grows up to 20 feet tall, has dark, glossy foliage and red fruit 5/8 inch or more in diameter, and is sometimes pear-shaped.

CRATAEGUS MOLLIS. The *scarlet hawthorn,* or *red haw,* or *downy hawthorn* grows to 25 feet, has hairy branches, slightly smaller red fruits, and does very well in poor soils.

Hawthorn flowers were often used to celebrate May Day in England and became a symbol of spring, often being called the *May tree* by poets.

JUNEBERRY (*Amelanchier* species). Sometimes called "the blueberry of the northern plains" and higher in vitamin C content than even citrus fruits, the juneberry was much used by various American Indian tribes who made it into the pemmican they carried with them on long journeys. The juneberry is also called *saskatoon* (in Canada), *serviceberry, sarvistree, May cherry, shadbush,* and *shadblow,* its last two names reflecting the belief that it blooms when the shad begin to run in spring. Its name *serviceberry* has a touching story behind it. Since its white blossoms bloomed almost as soon as the ground thawed in spring, pioneer families that had kept a body through winter to bury in workable ground used these first flowers to cover the grave. Long valued for its showy white blossoms, some species of juneberry can grow over 50 feet tall. All types have small, apple-shaped fruits that are bony inside but can be sweet and juicy (tasting like a combination of blueberries and cranberries) and even good to eat out of hand, although the fruit is usually made into jams, sauces, and pies. Fruit varies from berries the size of blueberries in some species to berries about as big as crabapples in others. All types are hardy from zone 2 or 3 southward, standing temperatures as low as −20° F., and make handsome ornamentals with

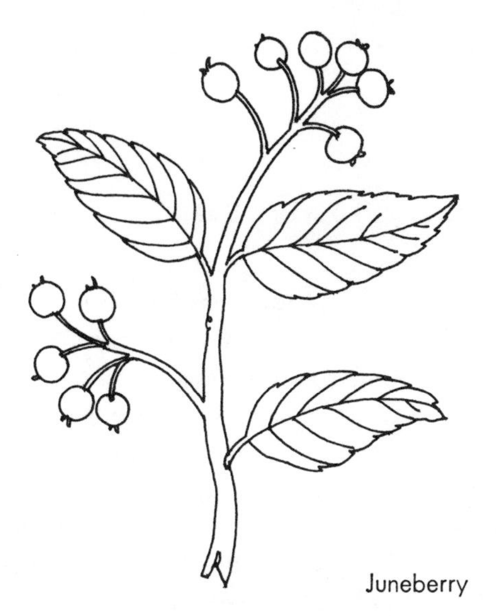

Juneberry

their profuse bloom, colorful fruit, and attractive bark and autumn foliage. The juneberry will do well in sun or shade, is of easy culture in any ordinary garden soil, and requires no spraying. It can be increased by sowing ripe seeds in spring or dividing clumps in autumn after the leaves have fallen.

Juneberries should not be planted within 500 yards of junipers as they are alternate hosts for some juniper rusts. Of the 25 or so species, all great favorites of birds, the following are the best to grow for their fruits.

AMELANCHIER ALNIFOLIA. The *saskatoon,* a shrubby Canadian species that grows 3 to 20 feet tall and bears sweet, juicy, black berries ½ inch in diameter and the best to eat of all. Available from Beaverlodge Nursery, Beaverlodge, Alberta, Canada; Field; Gurney. Named varieties include *Regent* and *Smoky.*

AMELANCHIER LAEVIS. A tree growing up to 30 feet high that is a beautiful ornamental and has sweet, purplish-black fruit. Offered by Raymond Nelson Nursery, Dubois, Pennsylvania 15801.

AMELANCHIER STOLONIFERA. Dwarf Juneberry or Quebec berry. A sprawling shrub about 3 feet high that spreads by means of underground suckers and often forms patches. Its sweet, juicy, purplish-black fruit is often used to make jelly. Sometimes sold by nurseries under the name Success.

AMELANCHIER OBLONGIFOLIA. The *Swamp sugar peas.* A 10 to 15-foot shrub with red-purple fruit once used to make "plum" puddings.

AMELANCHIER CANADENSIS. Called the *juneberry, serviceberry, shadblow.* The commonest juneberry. A 20 to 45-foot-tall tree with red-purple fruit that is relatively tasteless, but can be used for jam making or combined with other berries in pies. The Ute Indians, also called the Grasshopper Indians, actually made a "grasshopper fruitcake" with dried pulverized grasshoppers and juneberries. Following is a southern recipe more appetizing to most American palates:

Juneberry Hominy Salad

2 cups hominy
2 cups fresh juneberries
½ cup shredded coconut
½ cup chopped almonds
2 teaspoons honey
½ cup soy milk powder
⅔ cup vegetable oil
2 tablespoons lime juice

Combine hominy with juneberries, coconut, almonds, and one teaspoon of honey. Combine soy milk powder and one teaspoon of honey in a blender. Then blend at high speed, slowly adding oil until the mixture thickens. Stir in lime juice and fold the soy cream into the hominy mixture.

JUNIPERBERRY (*Juniperus* species). The berries of several junipers were used by the Indians for food and an aromatic oil is obtained from certain species. Probably the most famous of the juniperberries is that of the common juniper (*Juniperus communis*), which has been used for centuries to flavor gin and game dishes like wild boar. However, the western juniper or yellow cedar, a plant that can be a shrub or a tree 60 feet high, bears oval, blue-black berries with a bloom that are widely known to be edible and high in vitamin C—they were, in fact, used in the past as a scurvy preventive. Junipers are unusual in that the female flowers are made up of little scales which become fleshy, grow together, and form a sweetish, berrylike fruit. In certain species the small, pearl-shaped berries take three years to ripen. Following is an old bootlegger's recipe for Bathtub Gin made from juniperberries. I have not tested it, do not recommend it, and offer it merely as a historical curiosity that you must take full responsibility for if you make it or drink too much of it.

Bathtub Gin

2 parts alcohol
3 parts water
1 teaspoon juniperberry juice
1 tablespoon glycerin (to smooth it)

It cost bootleggers about two cents an ounce to make this concoction, which was ready to drink upon mixing.

KENTUCKY COFFEE TREE (*Gymnoclades diorca,* sometimes called *G. canadensis*). Through one source says this berry is "much too harmful to be used as a substitute for coffee," it certainly was so employed in colonial times and again during the Civil War. The tree was variously called the *Kentucky coffee tree,* the *coffee tree,* the *coffee berry,* the *Kentucky locust,* the *nicker tree,* and the *American mahogany,* taking the last name because its wood is a reddish color when cut and is used like mahogany in cabinet making. Resembling the locust or the black oak, this tall, handsome tree grows up to 90 feet tall and bears pods in which are enclosed flat brown seeds about the size of coffee beans. Hardy from zone 3 southward, it is easily grown in ordinary garden soils and is best propagated by root cuttings or by seed.

KILLARNEY STRAWBERRY (*Arbutus unedo*). No one knows how the Killarney strawberry, a native of the Mediterranean, spread to Ireland and the shores of Lake Killarney, though one ancient folk tale claims that it grew as a miracle for a Spanish monk longing for his home in Spain. At any rate, the *evergreen tree strawberry,* as it's also called, is an attractive ornamental up to 20 feet tall that bears juicy, orange-red, strawberrylike fruit about ¾ inch in diameter. Unfortunately, these strawberries that grow on trees are flavorless, but they can be made into a nice jam or wine. Killarney strawberries have survived temperatures of −20° F., yet they are not recommended for areas farther north than zone 5 because they will not fruit well in colder climates. The tree strawberry thrives in light, well-drained, sandy or peaty soil, a location sheltered from wind, and does not like a humid climate. It can be propagated by seed sown under glass in March, or by cuttings of partially ripe wood taken in the fall and grown under glass. The Killarney strawberry may not be a great fruit to eat, but it is a great curiosity. Its white flowers, which look like lily of the valley, take a full year to develop into fruit, ripening in the autumn just when the tree is flowering again—so that flowers and fruit are on the tree at the same time!

KIWI (*Actinidia sinensis*). The *Chinese gooseberry,* native to China's Yangtse Valley, was dubbed the kiwi because most of the fruit sold in American and European markets is grown in New Zealand. The kiwi is a many-seeded berry that tastes something like a gooseberry but really has its own delicious unusual flavor—like a combination of bananas and peaches, to my taste. Peeled of its woody skin and served sliced thin it makes an excellent dessert fruit. Kiwis grow to about 2 inches in diameter on a twining vine that climbs up to 30 feet high and does best in rich, loamy soil, although it isn't finicky about soil conditions. The plant does require full sun for good-flavored fruit and is hardy in America only from zone 6 southward, though it can be grown in zone 5 in protected places. A pretty ornamental vine with deciduous heart-shaped leaves that completely covers arbors and trellises, it has whitish flowers and hairy, reddish-brown, globular fruits resembling gooseberries that are green-colored inside with many seeds. It can easily be propagated by seed sown in early spring, by cuttings of partly ripened wood made in summer, or by layering. Another name for the fruit is the *Yangtao*. Relatives that aren't as well-known but are often just as tasty include the following, all of which are unisexual—requiring planting of both a male and female plant if they are to bear fruit:

ACTINIDIA ARGUTA. The *tara vine* or *Siberian gooseberry*. A higher-climbing vine that yields smaller, sweet, yellowish fruits.

ACTINIDIA KOLOMIKTA. The *Kolomikta vine* or *Manchurian gooseberry*. A handsome, smaller vine, rarely climbing up to 10 feet, that also yields small, sweet, yellow fruit.

ACTINIDIA CORIACEA. The *Chinese egg gooseberry*. Small, brownish, egg-shaped fruits speckled with white on a vine that grows 20 to 25 feet tall.

ACTINIDIA PURPUREA. The *purple Chinese gooseberry*. Yields smallish purple fruits on a vine growing about 25 feet high.

ACTINIDIA VOLUBILIS. The twining *Chinese gooseberry*. Brownish fruits on a vigorous climber that grows up to 30 feet.

ACTINIDIA POLYGAMA. The *silver vine,* or *cat plant*. Yields small, yellowish fruit on a 15-foot vine with attractive, silvery-white leaves (especially on the male plants, which bear no fruit). For some unknown reason cats are attracted to this plant and it has to be protected from them until established.

LEMONADE BERRY (*Rhus trilobata*). Various sumacs are used to make refreshing drinks that taste very similar to lemonade. The *lemonade berry* or *lemonade sumac* or *lemita* (from the Spanish *limonita*) is the most valued of them. A many-stemmed, 18 to 36-inch-high, deciduous shrub with ill-scented, coarsely toothed, trifolate leaflets and greenish flowers in small spikes that bloom before the leaves expand, it bears upright clusters of attractive, hairy, red berries. Also called *three-leaved sumac,* because of its three leaflets, the lemonade bush is hardy from zone 3 southward, grows in almost any garden soil, even in dry or rocky ground, and is easily raised from seed. Its berries should be gathered during dry spells for the highest flavor. Other close relations used for the same purpose are the *evergreen sourberry* (*Rhus integrifolia*) of southern California; the *evergreen sugar bush* (*Rhus ovata*) of southwestern U.S. desert regions, which is also called the *lemonade* and *sugar tree;* and the *smooth sumac* (*Rhus glabra*). All are beautiful, fall foliage plants for the garden, turning a more brilliant red than even dogwoods. *Poisonous sumac* (*Rhus Vernix*) is one of the two serious contact poison plants in the United States (the other is, of course, poison ivy). It is a tall shrub with reddish twigs found in swampy places, with 7 to 13 leaflets and white fruits instead of red. Incidentally, although poison sumac isn't cultivated, the plant's juice has been used as a lacquer. Still other sumacs have been used medicinally and sumac leaves are often gathered because they are rich in the tannin so important to the tanning and dyeing industries (the USDA will send you information on how to make money collecting the leaves). Lemonade made from crushing sumac berries in water has to be strained through cheesecloth to eliminate the fine hairs from the berries. The amount of sugar that must be added depends on the species used.

MYRTLE (*Myrtus communis*). Ovid wrote about this reputedly aphrodisiac berry in *The Art of Love:*

She was standing with her locks wreathed with myrtle. She gave me a leaf and a few berries. Recovering them, I was sensible of the divine influence as well; the sky shone with greater brightness and all cares departed from my breast.

Myrtle no doubt has the greatest reputation of any aphrodisiac berry. The strawberry has been mentioned as a love berry and the sugarberry (see Index) has something of a scarlet past, but among the berries only myrtle and allspice have really been valued as love foods. Allspice, or pimento (*Pimenta officialis*), is the dried, unripe berry of an aromatic 20 to 40-foot West Indian tree and isn't much grown in the garden. Myrtle, however, is easy to grow and has more of an amorous reputation. Myrtle, which originated in western Asia, is believed by the Arabs to be one of the three things (along with a date seed and a grain of wheat) that Adam took with him when he was cast out of Paradise. Venus wore a garland of myrtle when she rose from the sea, according to Roman mythology, and when satyrs tried to watch her bathing in the nude, she hid behind a myrtle bush. Myrtle crowns were awarded to victors of the Greek Olympic games and the plant has been a symbol of strength and love since ancient times. The Romans offered myrtle to Priapus as tokens of their gratitude for success in sexual affairs and the ancient Britons dedicated the plant to their goddess of love, always including myrtle in bridal bouquets and often planting myrtles near the homes of newlyweds. Mentioned in Petronius' *Satyricon,* myrtle berries, leaves, and flowers were used in many love potions and the plant's aromatic leaves and flowers have long been employed in the perfumery.

Myrtle is widely planted outdoors in America throughout zones 7, 8, and 9 for its aromatic evergreen leaves, profuse sweet-scented white flowers, and half-inch-long bluish-black berries. Growing 3 to 15 feet high, it prefers a sheltered spot in the garden similar to the evergreen thickets where it is found in the wild. It isn't too particular about soil, thriving in dry, rocky places, and needs little pruning. The bush can be increased by seed sown in spring or summer, or by cuttings of half-ripened wood taken in August and raised under glass.

In northern areas myrtle is a common greenhouse plant, often grown by florists for decorations. If the plants are potted and brought in for the winter to a cool greenhouse or conservatory, they should be provided with a loamy, well-drained soil and pruned to an appropriate size.

There are well over 100 species of myrtle from both the Old and New Worlds, but only a few are of interest to the gardener. Besides the tree myrtle (*Myrtus communis*) there is a form with three leaves instead of two at every joint that is used by the Jews for religious ceremonies such as the Feast of Tabernacles. Another popular type is the ugni or Chilean guava (*Myrtus ugni*), a shrub or small tree not over 20 feet high that is hardy from zone 6 southward and can be grown in the greenhouse farther north. An attractive bush with shiny green leaves that are white beneath and rose-pink flowers, it bears purplish-red, pleasantly flavored, juicy berries the size of peas that make delicious jellies and jams. It, too, should be grown outdoors against a wall, or in another protected spot, and can be propagated by seed or cuttings.

The tree myrtle's berries were used by the Romans to make a sauce eaten with wild boar, and the Corsicans still make an aromatic liqueur from them. A condiment can also be made from myrtle berries, but the most interesting recipe for the plant is this intriguing medieval one recommended for "sluggish lovers":

> *The flower and leaves of myrtle two handfuls infuse in two quarts of spring water, and a quart of white wine for 24 hours and then distill them in a cold still and this will be a strong scent and tincture, and by adding more or less of the myrtle you may make it stronger or weaker as you please.*

This beautifies and mixed with cordial syrups is a good cordial and inclines those that drink it to be very amorous.

OREGON GRAPE (*Mahonia nervosa*). The state fruit of Oregon, comprising its floral emblem, was once marketed in America and is used to make jellies, jams, wine, and other drinks. In addition to eating its showy clusters of blue berries the Indians used the bark of the Oregon grape, or *holly-leaved barberry,* for a tonic tea and obtained a yellow dye from its wood. The plant, named for American horticulturist Bernard M'Mahon, is widely grown as an ornamental today, as an attractive, low-growing evergreen for the shrubbery border or foundation plantings. It requires a place in the garden sheltered from winter wind and sun and likes a well-drained, loamy soil, but needs little care once established. The Oregon grape is easily propagated by seed sown in spring, suckers, layers, and cuttings of half-ripe wood rooted in sandy peat under glass. There are a number of species to choose from, but the two best types for edible berries follow.

MAHONIA NERVOSA. The true *Oregon grape.* A 1 to 2-foot-high shrub sporting oval, leathery leaves with spiny teeth on their margins, bright, fragrant yellow flowers and oval, dark-blue fruit with a bloom that hang in clusters like grapes. Hardy from zone 4 southward.

MAHONIA AQUIFOLIUM. Also called the *Oregon grape,* but a taller shrub, 3 to 10 feet high, with similar leaves and flowers and smaller bluish berries. Hardy from zone 3 southward.

POKEWEED (*Phytolocca americana*). Make no mistake about it—the root (which looks like horseradish) and seeds of pokeweed *are violently poisonous and people have died from eating them.* So are pokeweed's leaves and stems poisonous after turning red in autumn. Pokeweed or poke, however, is prized by some for its tender young stalks, which are gathered soon after they emerge from the earth and cooked like asparagus. Sometimes the young leaves are used, too, after much boiling. What is of interest here, though, is that the poisonous berries (the *seeds* in them are poisonous) have been used in the past to make pokeberry ink, pokeberry dye, and even to make a medicine I'm sorry to say I can't recommend for anything. *Pokeweed should not be grown on the home grounds and children should be warned about its poisonous parts.* It is a strong-smelling, handsome plant that grows 6 to 10 feet tall, has oblong-oval leaves 6 to 9 inches long that are often red-veined or red-stalked, white flowers, and blackish-red berries. Don't be fooled by birds eating the berry, seeds and all—they are apparently immune to the poison. Stick to making ink from them as a curiosity.

ROSSBERRY. A creation of Luther Burbank that tastes like a combination of the blackberry, loganberry, and raspberry, the rossberry is a good bet for the Southwest, where both cold and drought are problems. A prolific plant (yielding up to 600 gallons an acre) that fruits a year after planting, it bears big berries that are the first to ripen in the Southwest. The plants can easily be trained on a trellis or other supports and have the same cultural requirements as loganberries (see Index), being relatively free of insect pests and diseases. This berry has very few seeds and is excellent for eating out of hand as well as for pies, preserves, and quick freezing. It is available from Wolfe Nursery, Highway 377 West, Stephenville, Texas 76401.

ROWANBERRY (*Sorbus ancarparia edulis*). The European mountain ash, as the rowanberry is also called, has an interesting history. The *rowan tree* takes its name from the Danish "rune," meaning magic, and it was so called because it was supposed to have magic powers to ward off evil spirits. Often planted as an ornamental today, the rowanberry grows 50 feet tall and the *edulis* variety produces small, orange, berrylike pomes that are used in making jelly, jam, and wine. Rowanberries can be propagated by cuttings or seeds, which take two years to germinate, and are hardy from zone 2 southward. The Finnish whitebeam (*Sorbus hybrida*) is a smaller tree, growing from 15 to 40 feet tall, that is valued for its light red berries, these also used in jam making.

RUSSIAN OLIVE (*Elaeagnus angustifolia*). Sometimes called the *oleaster* or the *Trebizond date,* this 10 to 20-foot-high Eurasian tree with spiny branches and silvery leaves is widely grown as an ornamental and as a windbreak. It bears a yellowish, egg-shaped, berrylike fruit (not a true berry) that is about ½ inch long and tastes sweet but mealy. The flowers producing the fruit are very fragrant, so much so that the Russians once believed it excited women sexually and wouldn't allow their wives outside when the tree was in bloom. The Russian olive tolerates a variety of dry sites, and is hardy from zone 2 southward. It is easy to propagate by layers, root cuttings, or stratified seeds sown the second year. Some forty species of *Elaeagnus* are known, including the following, which are also valued for their fruit.

ELAEAGNUS COMMUTATA. The silverberry or wolfberry. A handsome shrub with silvery leaves, fragrant silvery-yellow flowers, and sweet, mealy, silvery fruit about ⅓ inch long. The erect shrub grows 8 to 12 feet high, spreads by stolons, and is hardy everywhere.

ELAEAGNUS MULTIFLORA. The gumi, as it's called, is a spreading shrub 4 to 9 feet tall with yellowish-white silvery flowers. Its reddish-orange fruits, about ¾ inch long, are speckled with white dots and have a pleasant acid flavor. Long cultivated for its fruit in Japan, it is used for jams, jellies, pies, and tarts. Hardy from zone 3 southward.

SALMONBERRY (*Rubus parviflorus*). Indians on the Pacific coast used these berries in a magic ceremony along with salmon; hence the berry's name. The Indians also ate the reddish-brown berries, which are a form of raspberry and a great favorite of birds.

SCOOTBERRY (*Streptopus roseus*). The "scoots" was slang for diarrhea in nineteenth-century New Hampshire and since these sweetish red berries always acted as a physic on the youngsters who eagerly ate them, they were called "scoot" berries. The scootberry grows to 2½ feet tall, has a creeping, much-branched rootstock and hairy, slightly branched stems. Its leaves are ovalish or lance-shaped, partly clasping the stems. The plant's purplish-pink, bell-shaped flowers grow on a short, twisted stalk from the axils of the leaves, giving it its other common name *twisted stalk.* Also called *liverberry,* it is native to eastern North America, preferring damp, shady places and an acid soil of pH4 to 5. It is increased by division of the rootstocks.

SEA BUCKTHORN (*Hippophae rhamnoides*). The Germans valued sea buckthorn or the *seaberry* so highly for its vitamin C content during World War II that they placed it on a list of plants to be protected as essential to the war effort. Orange-colored sea buckthorn berries have historically been used for making jellies, marmalades, and sauces, including a jelly that is served as a relish with fish. The spiny shrub or small tree, native to Eurasia, grows from 10 to 25 feet tall and is an attractive ornamental with its deciduous grayish-green leaves, silvery on the underside, and its colorful berries persistent most of the winter. The bush takes its name from the fact that it likes to grow near the sea, even on shifting sand, and from its many spines or thorns. Nevertheless, if planted out in the open where it has plenty of light, it will do well in any ordinary garden soil. To ensure fruit, both male and female trees must be planted within close range of each other, one male to every six or seven females. The bushes need no special care and are easily propagated by seeds, cuttings, or layers. Sea buckthorn also increases from suckers that can quadruple the number of bushes in an area within 4 to 5 years. The berries, technically small nuts enclosed in juicy flesh, are up to ⅜ inch long and vary from round to almost egg-shaped.

SOAPBERRY (*Sapindus* species). The scientific name for this tree explains the berry's use—*Sapindus* is a combination of the Latin for soap and "Indus" (Indian) in reference to American Indians using the berries for soap. Soapberries, the pulp of which contains saponin, lather up easily and were valued for shampooing, although the soap made from them does damage some materials. Two species are of horticultural importance. *Sapindus marginatus* is a deciduous tree up to 30 feet tall with yellow, egg-shaped fruit about one inch long that grows only from the southern part of zone 7 southward. *Sapindus Saponaric,* an evergreen, also grows up to 30 feet, but is only hardy in zone 9—that is, southernmost Florida. Both trees can be propagated by seed and do best in dry, sandy soil. The lather-producing agent saponin in soapberries can be poisonous if taken internally; in fact, American Indians caught fish by stupefying them with bits of the fruit thrown into pools.

STRAWBERRY-RASPBERRY (*Rubus illecebrosus*). Also called the *balloon berry,* because it is rather big—up to one inch long—the strawberry-raspberry is no good to eat out of hand, tasting sour or insipid, but has the flavor of mixed raspberries and strawberries when cooked and made into jams, syrups, and pies. Native to China, the plant is hardy here from zone 3 southward. An arching, very prickly plant about one foot high, it is cultivated like raspberries, but is very easy to grow, prolific of red berries, and has no special soil requirements. It makes a good ornamental ground cover for a bank and is best propagated by the division of roots in the fall.

SUGARBERRY (*Celtis* species). Sugarberries may have been the magical food of the Lotophagi, or lotus eaters, those legendary people who lived on the northeast African coast and gave wine made from the purplish-green fruit to travelers, who lost all desire to return home after drinking it (another possibility is the Chinese jujube). In any event, the berries are sweet and good for desserts as well as eating out of hand. *Celtis australis,* the 40 to 70-foot-tall tree the lotus eaters might have grown, is a Eurasian

Sugarberry

and North African tree little known outside of California in America, but the two attractive species below are more extensively cultivated here. Both are spiny trees that look like elms, have sweet fruit, and are easy to grow in any common garden soil. They are propagated from seeds sown in the fall, or by cuttings taken in the fall.

CELTIS LAEVIGATA. Growing 50 to 90 feet high, the *sugarberry* has 2½-inch, oblong leaves without marginal teeth, and fruit that is egg-shaped, ⅓ inch in diameter, and black-purple when fully ripe (at first orange-red). Hardy from zone 3 southward.

CELTIS OCCIDENTALIS. The common *hackberry, nettle tree,* or *hog berry* grows up to 100 feet in height and has toothed leaves except at the base. It bears somewhat pear-shaped fruits about ½ inch long that are black-purple when ripe. Hardy from zone 2 southward, it stands poor soil and smoke very well.

TWINBERRY (*Mitchella repens*). Also called the *partridge berry, squeakberry,* and *teaberry,* this trailing evergreen native to eastern North America makes an excellent dense ground cover for shady places in the garden if it is provided with a rich woods soil. Named for Virginia botanist Dr. John Mitchell, the plant, with its heart-shaped leaves, is a sparse producer of bright red, spicy berries that are good for eating out of hand. The berries, each one produced from unique twin flowers united at their base, are only about ¼ inch in diameter and are actually drupes with 8 nutlets. Twinberry is a good plant for a terrarium. It roots easily at the joints and can be propagated by layering or by division of its roots.

UNUSUAL TOMATOES (Solanaceae family). As noted, tomatoes are technically berries, but since covering all the common garden varieties would take a book which I've already written (*The Great American Tomato Book,* Doubleday, 1977), only several of the more unusual, little-grown types are briefly included here. Unlike garden tomatoes, most of these are used to make jams, preserves, and desserts.

CYPHOMANDRA BETACEA. The *tree tomato.* A sweet, subacid fruit, it can be eaten raw but is best when cooked for jams and preserves. A very ornamental tree that was first grown by the Indians of ancient Peru on mountainsides at elevations up to 8,000 feet, the tree tomato grows up to 12 feet tall, has foot-long elephant-eared leaves, fragrant purple and green flowers, and yields clusters of fruits that are egg-shaped, smooth-skinned, and orange-red when ripe. It bears fruit up to seven months a year and can be grown outdoors where the temperature never falls much below 50° F., but must be taken in during the winter months elsewhere and grown in a greenhouse or a large, sunny window. Seedling plants grown in containers should be potted in rich loam with leaf mold and well-rotted manure added to it, the size of the pot increased every year to accommodate root growth until the plant finally occupies a large pot. The plants are usually cut back in winter to encourage spring flowering, that being the extent of all pruning. Tree tomatoes are easily grown from seed but potted plants can also be purchased. A source for both is Lakeland Nurseries.

PHYSALIS PRUINOSA. The *ground cherry, strawberry tomato,* or *husk tomato.* This plant resembles a dwarf cherry-tomato bush, 18 to 24 inches tall, but bears deep yellow, cherry-sized fruit that is all wrapped up in a parchmentlike husk. The famed Poha Jam of Hawaii is made from husk tomatoes and the fruits are also dried in sugar and used like raisins. The ground cherry is an annual that is grown exactly like garden tomatoes and bears prolifically. Sources for this and the husk tomatoes following are Burgess, Farmer, Jung, and Gurney.

PHYSALIS IXOCARPA. The *tomatillo, Mexican ground cherry, jamberry.* Though it is a taller plant, 3 to 4 feet high, and yields larger fruit, up to an inch in diameter, the tomatillo is cultivated exactly the same as the ground cherry above and used in the same way. The tomatillo fruit, however, completely fills its husk and often bursts, revealing the ripe fruit; thus it is more difficult to remove the sticky fruit from the husk, though this can be done by soaking the husk in water.

PHYSALIS PERUVIANA. The *Peruvian cherry* or *Cape gooseberry.* Another husk tomato, but one that is treated as a tender perennial; that is, it can be grown outdoors when temperatures don't fall below 45° F., or grown inside in a greenhouse or sunny window during the winter and transplanted outside again in the spring. Otherwise it should be cultivated the same as the above husk tomatoes. It bears cherry-sized fruits that are easily removed from the deep-gold husks.

THE GOLDEN BERRY. An interrelated cross of forms of *Physalis* that is the newest kind of husk tomato, first introduced in America by England's Thompson & Morgan Inc., in 1976. Golden Berry is cultivated exactly like other husk tomatoes, but yields up to four pounds of juicy-sweet fruit a plant.

WAX MYRTLE (see Bayberries, in the Index).

WINEBERRY (*Rubus phoenicolasius*). Although a native of China and Japan, the wineberry grows wild here, an escape from old gardens in the eastern United States that is now found from zone 3 southward. The fruit isn't as sweet as other raspberries and is more seedy. On the plus side, it is juicier and has the advantage of ripening all at once—shortly after the calyx lobes covering the berries open to expose them. Unlike other raspberries, wineberries are sticky to the touch, but their hairy canes are a reddish-

Wineberry

brown color, making them a nice ornamental. They are trained and cared for like raspberries and suffer less trouble from insects and diseases because of their calyx coverings. Wineberries make delicious jams and jellies, and children like to eat them out of hand. Don't let them go, however, or they'll form impenetrable thickets.

WINTERGREEN (*Gaultheria procumbens*). Here's a berry appropriate to end with—one that can be used like an after-dinner mint. The folk names of this native American plant are many, including: *wintergreen, creeping wintergreen, checkerberry, hillberry, cowberry, nannyberry, teaberry, mountain tea, Canada tea, ground berry, ground holly, spice berry, boxberry,* and *partridge berry.* A prostrate evergreen that grows only 2 to 6 inches high and spreads by means of underground suckers, wintergreen is hardy everywhere. *Gaultheria* is named in honor of a Canadian doctor, Pierre Gaulthier. Though it is commonly grown as an attractive ground cover, its nutritional value is often overlooked today. The scarlet, pea-sized berries make good breath-fresheners eaten out of hand, are used to make a liqueur, and, in addition to their spicy, aromatic taste, are said to have tonic properties. The plant's leaves are a source of wintergreen oil (now obtained synthetically by most commercial manufacturers) and were made into a refreshing tea by early settlers, who simply steeped the leaves in boiling water several minutes. Wintergreen is easy to grow, once given a partially shaded location and acid soil. It is difficult to transplant from the wild, however, and it is best to use potted plants or propagate by cuttings of half-ripe wood, the seeds being very small and hard to handle. Several other species of wintergreen are valued for their fruit as well:

GAULTHERIA HISPIDULA. The *Creeping snowberry, moxieberry,* or *running birch.* A creeping evergreen plant with white flowers and white fruits about ⅕ inch in diameter that have a birchlike aroma. A dessert delicacy that tastes like wintergreen but is a little acid.

GAULTHERIA SHALLON. Called *salal* or *salalberry* on the West Coast, where it is found in the wild, this winterberry species is a spreading evergreen shrub up to 18 inches tall, whose long-lasting foliage is often used for floral arrangements and Christmas decorations. Its juicy, purplish-black berries, about ½ inch in diameter, are pleasant to eat out of hand and can be made into jelly.

GAULTHERIA MIQUELIANA. The *miquel berry* is a small evergreen that rarely grows over 12 inches high. A native of Japan, it bears pea-sized white fruit that makes fair eating.

GAULTHERIA OVATIFOLIA. The *mountain checkerberry.* Another spreading evergreen shrub that grows only about a foot high. Found in the wild in the Pacific Northwest, it bears small, red, spicy berries.

APPENDIX

I

For the Birds

Birds of many feathers are probably the worst pests you'll encounter in raising berries, though you'll attract many beautiful species you wouldn't have seen otherwise. To prevent birds from devouring the entire crop try any of the following methods in addition to the time-honored scarecrow.

• Pieces of rope or garden hose placed at strategic sites are sometimes mistaken for snakes by birds, who keep their distances.

• White string wrapped around bushes appears like spiderwebs to birds and scares them off.

• Broken mirrors or aluminum pie plates strung on bushes often frighten birds away.

• Some gardeners leave transistor radios on in the berry patch, the birds appreciating neither rock nor classical music.

• Commercial noisemakers are also available, as are recordings of bird distress calls.

• Volch oil sprayed on berries keeps birds away from the ripe fruit.

• Another ploy is to plant berries for the birds; that is, plant enough of their favorites like mulberries to divert them from precious raspberries and strawberries, etc. Birds have actually been known to get drunk on overripe, fermented, wild berries. It is not unusual for waxwings, for example, to get smashed on fermented rowanberries and crash into cars on the road, and ducks can get so drunk on overripe mulberries that they cannot fly.

• All berries can be protected by covering them with cheesecloth or clear plastic. Either can be draped around individual bushes or spread over the strawberry patch. Cheesecloth, because it comes in narrow widths, has to be sewn together to make a piece big enough to cover a strawberry patch. It can be spread right over the plants and anchored down on the sides or attached to a frame constructed around the bed. Clear plastic can be used in the same ways, but punch small holes in it to provide good air circulation for the plants.

APPENDIX

II

Fifty Mulches for Berries

ALUMINUM FOIL- repels some insects, and reflected light from it often increases yields. Weight down with stones around plants or cover with a heavier, more attractive mulch.

ASPHALT PAPER- forms a long-lasting barrier against weeds, and can be covered with a more attractive organic mulch.

BAGASSE (Chopped Sugar Cane)- long-lasting, clean, light-colored, and holds water well. Apply two inches deep. Sometimes sold as chicken litter.

BARK, SHREDDED- lasts a long time, is aesthetically pleasing, adds much humus to the soil, retains moisture, and will not blow away. Apply two inches thick.

BLACK PLASTIC- see Plastic Film.

BUCKWHEAT HULLS- dark, attractive, long-lasting, and do not mat—but blow away when dry. Apply three inches thick.

CLOTH- burlap, old rags, old rugs, etc., can be laid between rows if appearance is not important.

COCOA-BEAN HULLS- tend to pack and mold; have a chocolate odor for a few weeks; best mixed with other mulches. Apply two inches thick.

COCONUT FIBER- hard to get, but is long-lasting and attractive.

COFFEE GROUNDS- an excellent soil conditioner with attractive color, but is slightly acid. Sprinkle a little lime on them.

COMPOST- use when only half rotted; cover with other mulches.

CORNCOBS, GROUND- need a little nitrogenous fertilizer sprinkled over them; may attract vermin. Apply two inches thick.

CORNSTALKS- can either be shredded or used whole, covered with more attractive mulches.

COTTONSEED HULLS- good, but hard to obtain.

DUST MULCH- this simply means shallow cultivation of the soil to create a layer of dust that prevents upward movement of water and thus reduces evaporation. Some experts say all it really does is kill weeds by the act of cultivating.

EXCELSIOR- a good mulch, long-lasting, nonpacking, and weed free, but highly flammable. Apply two inches thick.

FIBER-GLASS MATTING- repels insects, is permeable to air and water. A brand called Weed Chek is widely available.

GLASS WOOL- good but tends to blow away unless covered with chicken wire or other mulch. Apply two inches thick.

GRASS CLIPPINGS- tests show this is one of the best mulches to repel nematodes and increase yield; however, grass clippings will mat, ferment, and smell badly if used fresh, and can harbor insects. Use *dry* grass clippings three inches thick or mixed with other mulches; if you use the dry clippings alone, try adding a half cup of blood meal per bushel.

GRAVEL, MARBLE, AND QUARTZ CHIPS- good on their own or for holding down other mulches, and protect well against mice. Experiments at the Colorado Agricultural Experiment Station showed that tomatoes mulched with black gravel yielded 10.27 tons per acre, while white gravel-mulched plants yielded 8.8 tons, and unmulched plants yielded 2.86 tons.

GREEN MANURE- cover crops, usually legumes, grown in place and cut for mulch.

HAY- excellent but may need fastening down. Don't use hay going to seed. Apply six inches thick.

LEAF MOLD- rotted leaves, especially nutritious for soil.

LEAVES- oak leaves are the best; avoid types like maple that tend to pack unless they are shredded first. You can dust oak leaves with a little lime if you like, but tests show their acidity doesn't affect the pH of the soil when used as a mulch. Wet down and apply one to two inches thick.

LICORICE ROOT- attractive, nonpacking, and long-lasting, but flammable and hard to obtain.

MUSHROOM SOIL, SPENT- has good color, benefits garden soil.

NEWSPAPERS- the newsprint repels insects and the wood pulp fertilizes plants, but newspapers can be unattractive and can form a tight mat so that air doesn't reach plant roots. Use four to six thicknesses around plants and cover with a more decorative organic mulch that will also help break the paper down. Newspaper ashes can be used as a mulch too.

PEANUT HULLS- excellent, but often blow away and may attract vermin.

PEAT MOSS- dries out and crusts too easily. Must be kept wet or it won't permit water to reach plants and will even suck water from the soil below to meet its own needs. Peat moss is best incorporated into the soil, but if it is used as a mulch, choose chunky types and keep them stirred up and wet. Not recommended for areas receiving little rainfall if rain is the only source of water.

PECAN HULLS- good mulch but hard to obtain.

PERLITE- blows away too easily, must be weighted down with another mulch. Yields no nutrients.

PINE NEEDLES- attractive and useful but flammable.

PLASTIC FILM (Polyethylene)- unattractive but very effective. Black (or green) plastic is preferred because it doesn't permit weed growth like clear plastic. Black plastic also absorbs the sun's heat during the day more than organic mulches do and radiates the heat back faster at night; thus plants mulched with it are less liable to frost injury than those mulched with organic materials. Its chief disadvantage is that it doesn't improve the soil any. Cut holes for your plants after it has been laid down as a mulch. The procedure for laying down black plastic is simple. On a windless day soak the ground thoroughly. Mark off the area to be covered by the plastic, dig furrows four inches deep along the edges of this space, unroll the plastic, and anchor it in the furrow with soil. Then make holes in the plastic and set in your transplants. Perforate the plastic for aeration. Remove the sheet at the end of the growing season and use it again next year if it's in good condition. Black or green plastic is best in an .0015 thickness and is available from most garden-supply stores.

SALT-MARSH HAY- one of the most effective mulches because it contains no weed seed, doesn't mat down, and is light and airy. Can be used for many years as it doesn't break down quickly.

SAWDUST AND WOOD SHAVINGS- sawdust is a good mulch that is not toxic to plants as is often written; however, it does consume a lot of nitrogen in decomposing, depriving plants of this nutrient. Apply three inches thick and add either a half cup of blood meal or a cupful of nitrate of soda per bushel. Wood shavings should be treated the same way.

SEAWEED- if you live along the coast this is a superb, mineral-rich, growth-promoting mulch. Can be placed directly around plants or composted first.

STRAWS- particularly good are wheat and oat straws, which are coarser than hay but more durable. Straws, however, are more flammable than hay and contain weed seed. Apply four inches thick.

ROCKS AND STONES- attractive, warm up the soil, and add trace elements to the soil as they imperceptibly wear away. Ring each plant with them a half inch from the stem and five inches outward, piling the rocks three inches high. Round, flat rocks weighing a few pounds are the easiest to work with, but many rocks and stones can be used. "Stones" in a wide variety of shapes can even be made from cement.

VEGETABLE PEELINGS- good, generally, but don't use old tomatoes or potato peelings on plants—tomatoes used as a mulch can cause canker and bacterial spot; potatoes can introduce verticillium organisms into the garden. Compost these waste products instead so that heat will destroy any diseases.

VERMICULITE- see Perlite.

WATER- in recent USDA experiments, 6-mil clear plastic bags filled with water increased crop yields by up to 20 per cent. By absorbing intense heat from the summer sun at midday and reducing heat loss at night, a water mulch allows you to plant earlier in the spring and extend your planting into the late fall. Heated water in the bags warms up the soil even more.

WEEDS- if weeds haven't gone to seed, they can be used as a mulch. So that they don't root again after a heavy rain, put them atop four or five layers of newspapers, which is a good way to get rid of two "waste products" at the same time.

WOOD CHIPS- more attractive than sawdust or shavings, but need the same amount of nitrogen.

APPENDIX

III

Seventy-eight Berry Sources

Following are the addresses of nurseries recommended for the various berries throughout this book. Those specializing in certain berries are so noted, but all offer a fairly wide range of stock. If you cannot find a specific berry or variety among these nurseries, try checking the *Plant Buyer's Guide* published by the Massachusetts Horticultural Society. The New York State Fruit Testing Cooperative, the New Jersey Small Fruits Council, and the North American Fruit Explorers (not a nursery) are good sources for getting information about unusual berries. Suppliers nearest your home are, of course, best where there is a choice:

Ackerman Nurseries—Bridgman, Michigan 49106 (Raspberries)

Ahrens Strawberry Nursery—R.R. 1, Box 721, Huntingburg, Indiana 47542

W. F. Allen Co.—Salisbury, Maryland 21801 (Strawberries)

Alpenglow Gardens—13328 King George Highway, Surrey, B.C., Canada

A. G. Ammon—Box 488, Chatsworth, New Jersey 08019 (Blueberries)

Dale Barsham Nursery—Alman, Arkansas 72921 (Raspberries and blackberries)

Beaverlodge Nursery, Beaverlodge, Alberta, Canada

Blandy Experimental Farm—Bogen, Virginia 22620

Boatman's Nursery—South Maple Street, Bainbridge, Ohio 45612 (Rabbiteye blueberries and raspberries)

Bountiful Ridge Nurseries—Princess Anne, Maryland 21853 (Raspberries and strawberries)

Brecks of Boston—Breck Building, Boston, Massachusetts 02210

James W. Brittingham—2538 Ocean City Blvd., Salisbury, Maryland 21801 (Strawberries)

E. J. Bryan—Washburn, Wisconsin 54891 (Raspberries and blackberries)

Buntings' Nurseries, Inc.—Selbyville, Delaware 19975 (Strawberries)

Burgess Seed & Plant Co.—Galesburg, Michigan 49053

W. Atlee Burpee Co.—Philadelphia, Pennsylvania 19132

D. A. Byrd—Locata, Michigan 49063 (Blueberries)

Cedar Grove Nursery—Cove City, North Carolina 27231 (Rabbiteye blueberries)

Chapman Berry Farm—E. Leroy, Michigan 49655 (Strawberries)

The Conmore Co.—P.O. Box 534, Augusta, Arkansas 72006 (Strawberries)

Cumberland Valley Nurseries—McMinnville, Tennessee 37110

Emlong Nurseries—Stevensville, Michigan 49127

Farmer Seed & Nursery Co.—Faribault, Minnesota 55021

Earl Ferris Nursery—Hampton, Iowa 50441

Henry Field Seed & Nursery Co.—Shenandoah, Iowa 51601 (Raspberries and currants)

Finch's Rabbiteye Blueberry Nursery—Bailey, North Carolina 27807

Dean Foster Nurseries—Hartford, Michigan 49057 (Currants)

Foster Nursery Co.—Fredonia, New York 14063 (Gooseberries and currants)

French Nursery Co.—Clyde, Ohio 43410

Fruit Haven Nursery—Kaleva, Michigan 49645 (Strawberries, raspberries, and blackberries)

Galleta Bros. Blueberry Farms—Hammonton, New Jersey 08037 (Blueberries)

The Guilde of Strawberry Bank, Inc.—93 State St., Portsmouth, New Hampshire 03801 (Wild strawberries)

Gurney Seed & Nursery Co.—Yankton, South Dakota 57078 (Strawberries, raspberries, currants)

Hartman's Blueberry Plantation—Grand Junction, Michigan 49056 (Blueberries)

H. G. Hastings Co.—Box 44088, Atlanta, Georgia 30336 (Strawberries, rabbiteye blueberries, and raspberries)

Hillemayer Nurseries—Lexington, Kentucky 40500

Ideal Fruit Farm & Nursery—Stillwell, Oklahoma 74960

Inter-State Nurseries—Hamburg, Iowa 51640

J. W. Jung Seed Co.—Randolph, Wisconsin 53956 (Strawberries and currants)

Keefe Blueberry Plantation—Grand Junction, Michigan 49056 (Blueberries)

Kelly Bros. Nurseries—Dansville, New York 14437 (Currants)

Krider Nurseries—Middlebury, Indiana 46540

La Fayette Home Nursery—La Fayette, Illinois 61449

Lakeland Nurseries—Hanover, Pennsylvania 17331 (Chinese gooseberry)

Lewis Strawberry Nursery—Rocky Point, North Carolina 28457 (Strawberries)

Earl May Seed & Nursery Co.—Shenandoah, Iowa 51603

McFadden Seed Co.—P.O. Box 1600, Brandon, Manitoba, Canada (Currants)

Mich-O-Tenny Nurseries—Bainbridge, Ohio 45612

J. E. Miller Nurseries—Canadaigua, New York 14424 (Currants and raspberries)

Monroe Nursery Co.—Monroe, Michigan 48161

Mullins Plant Farms—Chattanooga, Tennessee 37411 (Strawberries)

Raymond Nelson Nursery, Dubois, Pennsylvania 15801

Neosho Nurseries—Neosho, Missouri 64850

New Jersey Small Fruits Council—P.O. Box 185, Hammonton, New Jersey 08037 (Strawberries)

New York State Fruit Testing Cooperative Association—Geneva, New York 14456 (Strawberries, raspberries, currants, and mulberries)

North American Fruit Explorers—1848 Jennings Drive, Madisonville, Kentucky 42431

Ozark Nursery—Tahlequah, Oklahoma 74464

Raynor Bros., Inc.—Salisbury, Maryland 21801 (Strawberries, raspberries, and blackberries)

River View Nursery—McMinnville, Tennessee 37110 (Blueberries)

Phil Robers Nursery—Lake Geneva, Wisconsin 53147 (Raspberries and blackberries)

Romines Plant Farm—Dayton, Tennessee 39321

H. B. Scammell & Son—Toms River, New Jersey 08753 (Blueberries)

Scarff's Nursery, Inc.—New Carlisle, Ohio 45344

R. H. Shumway—Rockford, Illinois 61100 (Raspberries and currants)

Smith Berry Gardens—Ooltewah, Tennessee 37363 (Strawberries)

Southmeadow Fruit Gardens—Birmingham, Michigan 48009 (Gooseberries and currants)

Spring Hill Nurseries—Tipp City, Ohio 45371 (Strawberries and raspberries)

Stark Brothers—Louisiana, Missouri 63353

Theodore Stegmaier Nursery—Route No. 4, Cumberland, Maryland 21502 (Raspberries and blackberries)

Sutton & Sons Ltd.—Reading, England

Tennessee Nursery Co.—Cleveland, Tennessee 37311 (Rabbiteye blueberries)

Thompson & Morgan Inc.—Ipswich, England

Vilmorin—4 Quai de la Mégisserie 75001 Paris, France (Strawberries)

Vitis Strawberry Farm—Box 552, Niles, Michigan 49120 (Strawberries)

White Flower Farm—Litchfield, Connecticut 06759 (Strawberries)

Wolfe Nursery—Highway 377 West, Stephenville, Texas 76401

Zilke Brothers—Baroda, Michigan 49101

Zollar-Greening—Benton Harbor, Michigan 49023 (Strawberries)

Index